Introductory Physics for the Life Sciences

This textbook provides an accessible introduction to physics for undergraduate students in the life sciences, including those majoring in all branches of biology, biochemistry, and psychology and students working on pre-professional programs such as pre-medical, pre-dental, and physical therapy. The text is geared for the algebra-based physics course, often named College Physics in the United States.

The order of topics studied in this volume requires students to first understand a concept, such as the conservation of energy, momentum, voltage, or current, the change in a quantity such as entropy, or the rules of ray and wave optics. Then, students apply these concepts to solve problems in the areas of thermodynamics, electrical circuit, optics, and atomic and nuclear physics. Throughout the text these quantity-based applications are used to understand systems that are critical to the understanding of biological systems, such as the entropy of evolution, the signal down the axon of a nerve cell, the optics of the eye, and the operation of a laser.

This is part 2 of a two-volume set; volume 1 introduced students to the methods of mechanics and applied these problem-solving techniques to explicitly biological topics such as the sedimentation rate of red blood cells in haemoglobin, the torques and forces on a bacterium employing a flagellum to propel itself through a viscous fluid, and the terminal velocity of a protein moving in a gel electrophoresis device.

Key features:

- Organized and centered around analysis techniques, not traditional mechanics and E&M.
- Presents a unified approach, in a different order, meaning that the same laboratories, equipment, and demonstrations can be used when teaching the course.
- Demonstrates to students that the analysis and concepts they are learning are critical to the understanding of biological systems.

David Guerra, Ph.D., is a Professor of Physics at Saint Anselm College (SAC), United States of America. During his time at SAC, he has taught many of the courses offered by the department, including the physics course for biology majors, and has developed and taught courses in such topics as laser physics and remote sensing. At SAC, he has conducted remote sensing research, both in instrument development and data analysis. Upon arriving at SAC, Professor Guerra designed, built, and operated a novel lidar (laser radar) that utilized a holographic optical element (hoe) as its transmitter and receiver. This work was done over the course of his first decade at SAC with several sets of student researchers and in coordination and with funding from NASA-Goddard Space Flight Center. The project resulted in a series of publications and the successful development of a lidar system that was flown by NASA using the (hoe) technology. In his time since the successful completion of the lidar work, Professor Guerra has focused his remote sensing research on the analysis of remote sensing data in the investigation of natural systems. As part of a National Science Foundation (NSF) grant, Professor Guerra conducted research with several sets of student researchers and faculty from other universities from across New Hampshire to study the environmental dynamics of the natural systems throughout their state. He continues his collaborative remote sensing work with faculty from the departments of computer science and biology at SAC, investigating relationships between environmental conditions and animal behavior and plant development. Professor Guerra has also done work in physics education research, ranging from the development of new laboratory experiences to new pedagogies and even chapters in a widely used high school physics textbook.

Introductory Physics for the Life Sciences
Volume 2
Quantity-Based Analysis

David V. Guerra

CRC Press
Taylor & Francis Group
Boca Raton London New York

CRC Press is an imprint of the
Taylor & Francis Group, an **informa** business

First edition published 2023
by CRC Press
4 Park Square, Milton Park, Abingdon, Oxon, OX14 4RN

and by CRC Press
6000 Broken Sound Parkway NW, Suite 300, Boca Raton, FL 33487-2742

CRC Press is an imprint of Informa UK Limited

British Library Cataloguing-in-Publication Data
A catalogue record for this book is available from the British Library

ISBN: 978-1-032-30041-2 (hbk)
ISBN: 978-1-032-31108-1 (pbk)
ISBN: 978-1-003-30807-2 (ebk)

DOI: 10.1201/9781003308072

Typeset in Times
by MPS Limited, Dehradun

Contents

VOLUME 2   Introduction

Volume 2

Introduction

This is the second of two volumes of an introductory, algebra-based textbook written for students interested in the life sciences. This volume is constructed around a quantity-based analysis that employs a set of rules associated with specific quantities in nature. Some of these quantities may sound familiar, such as energy, momentum, and angular momentum. In addition to the standard conserved quantities, there are other quantities, such as voltage, forms of thermal energy, entropy, and light, that stay constant or change in a prescribed way so they provide another way to analyze a physical system. It is this type of analysis, in contrast to mechanics, that is the consistent thread that is applied throughout this volume.

This quantity-based analysis is often used to study systems in which it is difficult to employ mechanics. For example, it is almost impossible to know all the forces on and the motion of every particle of a gas in a container as the gas is heated, or the forces on all the electrons as they travel down a wire, or even the forces on the ions as they transition and generate light in a laser material, so specific quantities and the problem-solving methods associated with these quantities are used to analyze these types of systems.

To provide mechanical context to the standard conserved quantities, the volume begins with an introduction of energy through work, momentum through impulse, and angular momentum through angular impulse. Once a dynamics-based understanding of these standard conserved quantities is established, the focus of this volume is the application of quantity-based techniques to the study of systems in which mechanics is not a good choice for analysis. These analysis techniques are consistent with the way physicists and engineers study these systems. For example, thermodynamics is based on the analysis of how thermal quantities like internal energy and entropy change when heat is applied to a system and not on the forces applied between particles as a material is heated. Electrical circuit analysis is based on the rules that govern the way voltage and current change in a circuit and not on the forces on the electrical charges in the wires. Optics is based on rules associated with the path of light rays or the addition and subtraction of waves and not on forces and motion. Finally, even through forces provide models for the study of atomic and nuclear systems, conserved quantities, such as energy, momentum, and angular momentum, provide the structure upon

DOI: 10.1201/9781003308072-16

which the analysis is built. Therefore, in this volume each section introduces the rules that govern and the analysis techniques that are applied in each branch of physics: thermodynamics, electrical circuit analysis, optics, and atomic & nuclear physics. Then, these rules are applied with a consistent focus on understanding and monitoring changes in these quantities as the systems evolve. It is this underlying method of analysis that is the structure upon which this volume is built.

16 Energy and Work

16.1 INTRODUCTION

Work and energy are the starting point of the study of conserved quantities and the analysis techniques associated with these types of quantities. Even though these terms, energy and work, are familiar words, their physical definitions may not be as well-known. Therefore, the chapter starts with a force-based definition of work and the subsequent definitions of potential and kinetic energy. Conservation of energy is introduced as the first of the conservation laws, and several examples of applying this law of physics are presented.

16.2 CHAPTER QUESTION

How is the length of the skid-marks left on a section of level road related to the speed of the car that made them? Specifically, are the skid-marks made by a car traveling at 40 mph more than twice, twice, or less than twice as long as those made by a car traveling at 20 mph? Also, does the weight of the car need to be known to make this comparison?

16.3 DEFINITIONS OF WORK AND ENERGY

The study of energy as a conserved quantity starts with the definition of work and energy. Unlike the concept of force, which we can understand through experience, the quantities of work and energy are more abstract. Work is defined as the product of a displacement and the components of the force that cause that displacement. For example, in Figure 16.1 a box is dragged across a level floor with a rope that is at an angle θ to the horizontal.

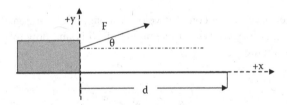

FIGURE 16.1 A box dragged across a level floor with a rope pulled at an angle θ to the horizontal.

The work (W) done on the box by the force, $\vec{F}$, associated with the displacement, $\vec{d}$, is given in equation (16.1) as,

$$W = F\, d\, cos\,(\theta_F - \theta_d) \tag{16.1}$$

Where F and d are the magnitudes of the force and displacement associated with that force, θ_F is the angle of the force, and θ_d is the angle of the displacement. In the diagram, the force is at the angle θ and the angle of the displacement is $0°$, so the angle in the argument of the cosine is θ. This is a common situation, since the direction of the displacement is often along the x-axis or y-axis, so θ_d is often $0°$. If the displacement and the force are both at angles relative to the +x-axis, then just find the difference and plug it into the cosine. Since work is the product of a force and a displacement, the

DOI: 10.1201/9781003308072-17

units of work are a combination of the units of force and displacement. Thus, in the metric system, the unit of work is the Joule (J), which is the product of a Newton (N) and meter (m), 1 J = 1 N m.

16.3.1   SOME NOTES ABOUT WORK

If there is no displacement in the direction of the force, there is no work, no matter how hard the force is applied. If you hold a heavy box for an hour but don't move, no matter how hard it may seem, you have done no physical work. Chemical reactions in your muscles may have exchanged energy, but no work was done. This is an important distinction to make as the types of work and energy are put together to study a situation.

There may be motion, but if the motion is not in the direction of the force, that force has done no work. So, if the displacement of an object is perpendicular to a force, the force did no work on the object. For example, gravity does no work on a satellite in a circular orbit about the earth, because the displacement is always perpendicular to the force of gravity. The normal force, which by definition is perpendicular to the surface, does no work on an object moving across a surface.

If work is done, it can be positive or negative depending on the relationship between the direction of the force and the displacement. For a box pulled across the floor, like the one in Figure 16.1, the work is positive, because there is a component of the force in the same direction as the displacement. For forces like the friction of a skidding bicycle, the bike will still be moving forward while the force is in the direction opposite to the motion. Thus, when an object is slowing down, there must be negative work.

Although work can be positive or negative, it is not a vector. It is a scalar, in which a positive sign indicates that the force had a component in the direction of the displacement and a negative sign indicates that the force had a component in a direction opposite the direction of the displacement. As indicated by the tense of all the sentences in this paragraph, work can only be measured for an event that has already been completed. Thus, work is always in the past and is referred to as an amount of work "done" by a force.

Example

A person pushes a crate forward at a constant velocity along a flat horizontal floor. The person is pushing with a force that is forward with a magnitude of 90 N and downward at 15.0° below the horizontal, as shown in Figure 16.2.

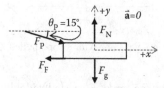

FIGURE 16.2   Work example.

During the time interval in which the force is applied, the crate moves 15 m forward at a constant velocity.

During that Time Interval:

    a. How much work is done by the person on the crate?
    b. How much work is done by the earth's gravitational field on the crate?
    c. How much work is done by the friction on the crate?
    d. How much work is done by the normal force on the crate?

Solution

a. The work done by the person is associated with F_P, so the force vector and the displacement vector are shown in Figure 16.3.

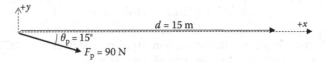

FIGURE 16.3 Work done by the person.

From the definition of work, notice that the angle of the force, θ_F, is 345° and not 15° because all angles are still measured for the +x-axis.

$$W_P = F_P d \cos(\theta) = F_P d \cos(\theta_{FP} - \theta_d) = 90N \cdot 15m \cdot \cos(345° - 0°) + 1304\ J$$

b. The work done by the earth's gravitational field is found from the force and displacement in Figure 16.4.

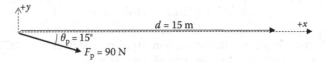

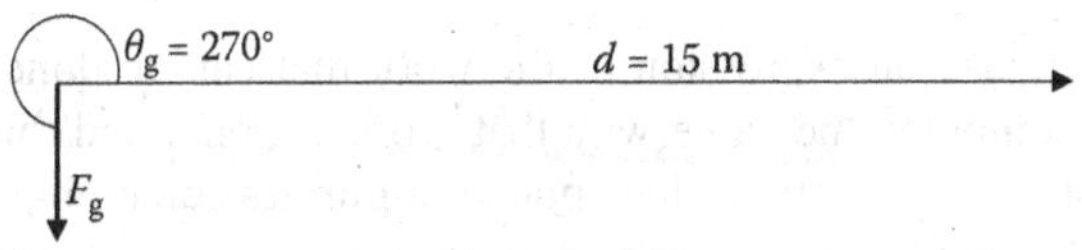

FIGURE 16.4 Work done by the earth's gravitational field.

From the definition of work:

$$W_g = F_g d \cos(\theta) = F_g d \cos(\theta_{Fg} - \theta_d) = F_g \cdot 15m \cdot \cos(270° - 0°) = 0\ J$$

The mass or the weight of the crate, F_g, is not given, but it doesn't matter because $\vec{F_g}$ is perpendicular to the displacement, so the work done by gravity on the crate is 0 J.

c. To find the work done by friction, the force of friction $\vec{F_f}$ must first be found:

Since the person pushes with a force F_P that is down and to the right at 15°, the x-component of this force is: $F_{PX} = F_P \cos\theta_P = 90\ N \cos(345°) = 86.93N$

Friction is in the opposite direction as the x-component of the applied force, so the x-component of the force of friction is: $F_{fx} = -F_f$.

Since the crate is moving at a constant speed, the x-component of the acceleration is zero. So, statics in the x direction gives: $F_{Px} + F_{f,x} = ma_x$ and $F_{Px} + (-F_f) = 0$. So, in conclusion,

$$F_f = F_{Px}$$
$$F_f = 86.93\ N$$

Work done by the frictional force on the crate with the displacement is found from the relationships given in Figure 16.5.

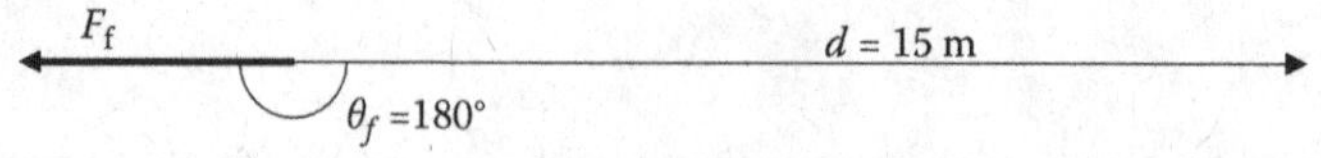

$$F_f \qquad\qquad\qquad\qquad d = 15\ m$$

$$\theta_f = 180°$$

FIGURE 16.5 Work done by the frictional force.

$$W_F = (86.93\ N)(15m)\cos(180°) = -1304\ J$$

d. There is a normal force on the crate, but it is perpendicular to the displacement ($\theta_{FN} = 90°$), so the work done by the normal force is zero.

$$W_N = F_N\ d\ cos(\theta_{FN} - \theta_d) = F_N\ d\ cos(90° - 0°) = 0$$

Calculating the total work done by all the forces acting on an object when that object undergoes a displacement produces the same result as calculating the net force acting on the object and then calculating the work done by the net force. Both ways have to yield one correct answer. For the example above, the net force is zero since the object has no acceleration, since it is moving at a constant velocity, so the net-work on the object by all the forces is zero. Note that the sum of the individual amounts of work done by the four forces in this example, 1304 J + 0 + 0 + −1304 J = 0 J, is also zero.

16.4 POTENTIAL ENERGY (U)

Given that work is in the past, an expression of the work that can be done in the future due to a specific arrangement is computed the same way that work is computed, but it is given the name *potential energy*. The name is appropriate since potential implies something that is possible, and in this case, it is the work that can be done. Since potential energy is just the work that can be done, it is computed with exactly the same equation as work is computed.

For some forces, the work done on a particle as it moves from point A to point B does not depend on the path it takes as it moves from A to B. These forces are called *conservative forces*. For conservative forces like gravity, electrostatic, and spring forces, a value can be assigned to each point in space, and the work done on a particle as it moves from any point A to another point B, as shown in Figure 16.6, can be computed.

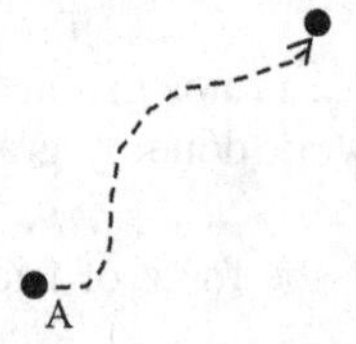

FIGURE 16.6 Work done from point A to point B.

The value assigned at points A and B is called the potential energy U, so that the work is the negative of the change in potential energy, which is demonstrated in equation (16.2) as

$$W_{AB} = -(U_B - U_A). \qquad\qquad (16.2)$$

The reason the change in potential energy is the negative of the work done is a temporal argument. If an object is displaced in a direction opposite the force on this object, the object, when released, will move in a direction back to where it began. An example, which will be illustrated in the next section, is lifting an object upward against the force of gravity.

16.4.1 Gravitational Potential Energy near the Earth's Surface (U_G)

As shown in Figure 16.7, when an object is located at height, h, above the ground, it has a force of gravity on it equal to the product of its mass, m, and the gravitational field strength, g.

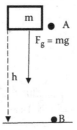

FIGURE 16.7 An object near the earth's surface.

If the object is dropped, it falls down a distance, h, from point A to point B. The gravitational field of the earth does work on the particle as long as point A and point B are not at the same elevation. Points A and B are arbitrary, so the reference level is commonly chosen as the bottom of the path, point B, so the height h is a positive value. Our assignment of a value of potential energy to each point in space is $U_g = mgh$, where h is how high above the reference level the point is. The work done by the gravitational field starts with equation (16.2) for a particle as it moves from point A to point B,

$$W_{AB} = -(U_B - U_A)$$
$$W_{AB} = -(mgh_B - mgh_A)$$
$$W_{AB} = -mg(h_B - h_A)$$

Notice if the object is starting at a height (h_A) and falling to a height (h_B), then $h_A > h_B$ so ($h_B - h_A$) is negative; therefore, gravity does positive work W_{AB} on the particle as it falls. The gravitational force is in the same direction as the displacement of a falling object.

Therefore, the gravitational potential energy at any point of an object above a set reference point is given in equation (16.3) as

$$U_g = mgh. \tag{16.3}$$

Example

An x-y coordinate system has been established for a region of space near the surface of the earth with +y being upward and the origin being at ground level. A particle of mass .20 kg moves from point A at (.2 m, 1.0 m) to point B at (.5 m, .4 m) (Figure 16.8).

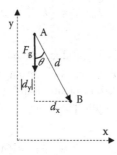

FIGURE 16.8 Gravitational potential energy example.

Find the work done on the particle by the gravitational field of the earth two ways:

 a. Calculate it as the force along the path times the length of the path, and
 b. Calculate it as the negative of the change in gravitational potential energy.

Solution

 a. Using the definition of work:

$$
\begin{aligned}
d_x &= x_B - x_A = .5m - .2m = .3m \\
|d_y| &= |y_B - y_A| = |.4m - 1.0m| = .6m \\
d &= \sqrt{d_x^2 + |d_y^2|} = \sqrt{(.3m)^2 + (.6m)^2} = .67082m \\
\theta &= tan^{-1}\left(\frac{d_x}{|d_y|}\right) = tan^{-1}\left(\frac{.3m}{|.6m|}\right) = 26.565°
\end{aligned}
$$

$$
W_{AB} = F_g \cos\theta\, d = mg\cos(\theta)\, d = (.2kg)\left(9.8\frac{N}{kg}\right)[\cos(26.565°)]\,(.67082\ m)
$$
$$
W_{AB} = 1.176\ J
$$

 b. We define our reference level to be at ground level ($y = 0$):

$$
\begin{aligned}
W_{AB} &= -(U_{gB} - U_{gA}) = -(mgh_B - mgh_A) = -(mgy_B - mgy_A) \\
W_{AB} &= mg(y_A - y_B) = (.2kg)9.8\frac{N}{kg}(1m - .4m) \\
W_{AB} &= 1.176\ J
\end{aligned}
$$

Notice how much easier solution b is compared to solution a. Keep this in mind as the chapter builds toward the concept of conservation of energy.

 Spring Potential Energy is stored when a spring is stretched or compressed a distance, $|s|$, from the equilibrium. The force provided by a spring when displacement $|s|$ is given by equation (14.2) is $F_{spring} = k|s|$, where k is the spring constant measured in N/m. So, as a spring is pulled or pushed away from its equilibrium position, the force is the opposite direction, and it gets stronger as the object is pulled further away. Since the force increases linearly, the average force across the entire displacement is just $\frac{1}{2}$ the force at the maximum displacement $|s|$, or

$$
F_{spring} = \frac{1}{2}k|s|,
$$

Therefore, the work done by the spring is the product of the force and the displacement for a spring stretched a distance s, which is $W_s = \frac{1}{2}k|s||s| = \frac{1}{2}k\,s^2$. From the mass on the end of the spring in Figure 16.9, if the 0 point is the equilibrium position and the spring is stretched to point A and released, it will move toward the equilibrium position.

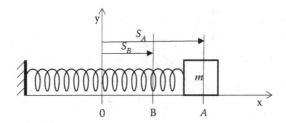

FIGURE 16.9 Mass attached to a spring with a constant k.

The work done by the spring will be $W_s = \frac{1}{2}k(s_A^2 - s_B^2)$. If this is substituted into equation (16.2), then the spring potential energy is given in equation (16.4) as

$$U_s = \frac{1}{2}k\,s^2. \tag{16.4}$$

This derivation makes sense, but it is not the complete story. To do the complete derivation of the work done by a spring on the block by the spring, use the force along the path times the length of the path. As the block moves from its initial position to its equilibrium position, the path needs to be divided up into an infinite set of infinitesimally small segments of force and displacement, each so small a stretch of the spring. Hence, the spring force can be considered constant while the block travels along that segment of the path. Calculate the work done on each path segment and then add up the infinite set of results. The process is known as integration, which results in the same form for equation (16.4), so this expression will be used as the spring potential energy.

16.4.2 FORCE VS. DISPLACEMENT GRAPHS AND WORK

If the force applied to an object is plotted (solid line) vs the displacement (dashed line) of the object, the area enclosed (diagonal line) by the force and displacement lines is the work done by this force over this displacement.

Example 1

Graph of weight vs. height is given in Figure 16.10.

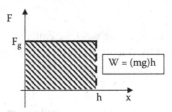

FIGURE 16.10 Graph of force vs. displacement for an object falling near the earth's surface.

Notice that the area enclosed by the force and displacement lines is the area of a rectangle with lengths of F_g and h. Since F_g is equal to the product of mass, m, and the gravitational field strength, g, the work done is just the product of force and displacement, *(mg)h*.

Example 2

Spring force vs compress distance.
    Graph of spring force vs. stretch length is given in Figure 16.11.

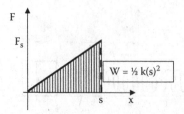

FIGURE 16.11 Graph of force vs. displacement for an object falling near the earth's surface.

Notice that the area enclosed by the force and displacement lines is the area of a triangle, which is $\frac{1}{2}\,bh$, where b is the base of the triangle and h is the height of the triangle. Since the base is s and the height is $F_s = k\,s$, the work done is $\frac{1}{2}ks^2$.

16.5   KINETIC ENERGY

The energy associated with the motion of an object is the *kinetic energy*. Since motion is either happening or not, at a particular instant of time, the kinetic energy is an expression of work in the present. The expression for kinetic energy must be equivalent to the expression for work, but must also include velocity. So, for an object of mass m starting at point A of Figure 16.6 and moving to point B, the work done on the object by a force that is in the same direction as the displacement is: $W = F\,d\,cos(\theta_F - \theta_d) = Fd$. If Newton's 2nd Law, is substituted into the expression of work through the force ($F = m\,a$) the expression of the work is: $W = (m\,a)\,d$. Next, equation (8.3), $v^2 - v_0^2 = 2\,a\,[d]$, from kinematics, is rearranged to the form: $a\,d = (1/2)(v^2 - v_0^2)$ and substituted into the expression for work and kinetic energy in equation (16.5) as

$$W_{AB} = \tfrac{1}{2}\,m\,(v_B^2 - v_A^2) = \tfrac{1}{2}\,mv_B^2 - \tfrac{1}{2}\,mv_A^2. \qquad (16.5)$$

The term associated with the velocity and mass of the object in equation (16.5) is the kinetic energy, which is assigned the variable K. Thus, the kinetic energy of an object with a mass m and moving with a velocity v is given in equation (16.6) as

$$K = \frac{1}{2}m\,v^2. \qquad (16.6)$$

The relationship between kinetic energy and work is called the **work-energy theorem**, and is generated by combining equations (16.5) and (16.6) to get equation (16.7) as

$$W_{AB} = K_B - K_A. \qquad (16.7)$$

In the expression for kinetic energy, the velocity is squared and mass is only positive, so the kinetic energy is always positive. This agrees with the concept; since kinetic energy is the energy of motion, an object is either moving or it is not. Remember, kinetic energy is energy in the present.

Example

Compute the kinetic energy in Joules of a 2 kg block moving at 4 m/s.

Solution

$K = \frac{1}{2}\,mv^2 = \frac{1}{2}\,(2\text{ kg})(4\text{ m/s})^2 = 16\,(\text{kg m/s}^2)(\text{m}) = 16\text{ Nm} = 16\text{ J}$

16.6 CONSERVATION OF ENERGY

The total Mechanical Energy (*ME*) of a system at a specific time "*i*" is the sum of the kinetic and potential energy at that instant of time and can be expressed in equation (16.8) as,

$$ME_i = (K_i + U_i). \tag{16.8}$$

Since, the quantities of potential energy and kinetic energy are all based on the concept of work and all three are the past, present, and future of the same quantity, then it is not difficult to understand that as time moves forward, there may be exchange of one form of energy for another, but the sum of these three forms of energy remain the same. This is the concept of conservation of energy, which states that the difference in the total mechanical energy at an initial moment (*i*) to the final moment (*f*) of a system is the work done (W_{done}) on the system:

$$W_{done} = ME_f - ME_i = \left(K_f + U_f\right) - (K_i + U_i)$$

It is common to rearrange this expression to provide the expression of conservation of energy as equation (16.9),

$$K_i + U_i + W_{done} = \left(K_f + U_f\right). \tag{16.9}$$

Conservation of energy states that the total sum of mechanical energy initially plus any work done, either positive or negative, during an event, is equal to the total mechanical energy after the event. Notice that while the total amount of energy is conserved, the distribution of energy may change. This is remarkable, considering energy is an abstract concept described from the concept of work. Time elapses, but the total energy stays the same; thus, it is said that energy is symmetric with time. It doesn't matter when the events occur, as long as all the energies are accounted for, the total is unchanged. There are many forms of energy – mechanical, chemical, electrostatic, heat, nuclear. Energy can be transformed from one kind to another, but the total amount of energy is constant. This remarkable concept of a conservation of energy is fundamental to our understanding of systems as varied as power plants, electrical circuits, and the structure of molecules, atoms, and nuclei.

Since work is defined as the product of a force and the displacement it causes, conservation of energy is used to analyze situations in which an object moves from one place to another. If work is positive, then the force is in the same direction as the displacement, so the objects will be moving faster or climbing higher. Conversely, if the work is negative, the force is in the opposite direction as the displacement, so the object will be moving slower or reaching a lower height. For example: When an object is dropped from rest at some height above the earth's surface, it starts with some U_g but no K. As the object falls toward the earth, it loses U_g and gains K. Just before the object hits the ground, it has lost all of its initial PE_g but gained an equal amount of K.

A concept map of conservation of energy is given in Figure 16.12.

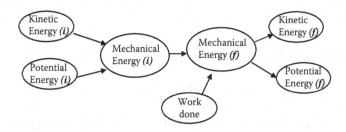

FIGURE 16.12   Concept map of conservation of energy.

This concept map is provided to emphasize the process used to analyze a system in this volume of the text. In this basic conservation of energy framework, an initial set of quantities, kinetic and potential energy, are computed. After an event, the same quantities and the work done during the event are computed and compared to the initial values. Upon comparison of the quantities, any unknowns are computed. This is a different procedure than the process of mechanics emphasized in volume 1 of this text.

In this specific case, the mechanical energy is initially comprised of kinetic and potential energy. The final energy is a result of these two energies. Any work done on the system also needs to be added to the final energy tally, since the energy of a system may change from one form to another, but the total stays the same, unless work is added.

It is critical to the process of solving problems with conservation of energy that energy and work are not double counted. It is most important to keep this in mind when trying to compute the work involved in these problems and others in the future. It is true that when an object is dropped from a height, the work done by gravity could be considered, but instead this will be accounted for with gravitational potential energy. So, the rule is, that work will be used in the conservation of energy problems to represent the effects of non-conservative forces like friction. So, to start, if a problem states that the system is frictionless, set the work done to zero and proceed to find the kinetic and potential energy terms.

16.6.1 CONSERVATION OF ENERGY EXAMPLES

Example 1

A block is lifted vertically upward to a height of 0.46 m above the ground and then dropped from rest. Assume air friction is negligible.

    a. At what speed will the block strike the ground?
    b. How much higher must the block be lifted before it is dropped so that it will strike the ground with twice the speed as it hit the ground in part a?

Solution

    a. Step 1. Generate a simple sketch to define the two instants of time i and f, as shown in Figure 16.13.

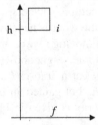

FIGURE 16.13 Example 1 sketch.

Step 2. Apply conservation of energy:

$$K_i + U_i + W_{done} = K_f + U_f$$

Since the block is at rest at the top (i) the $K_i = 0$ J. The U_i at the top is gravitational (mgh), and the work is 0 J, since the question specifies neglecting air friction. At the bottom, the kinetic energy is unknown, so insert the expression $\frac{1}{2} mv^2$, and the gravitational potential energy is 0 J, since this is the reference point of the height measurement. This results in a conservation of energy expression of: $0\,J + mgh + 0\,J = \frac{1}{2}\,mv^2 + 0\,J + 0J$. Dropping all the zeros gives: $mgh = \frac{1}{2}mv^2$

Since the mass is on each side, it can be cancelled from both sides of the equation, $gh = \frac{1}{2}v^2$

Rearranging and solving for v gives: $v = \sqrt{2gh} = \sqrt{2\left(9.8\frac{N}{kg}\right)(0.46\ m)} = 3\frac{m}{s}$

The units do work out to be meters per second, since through Newton's 2nd Law a Newton (N) is equivalent to $N = kg\frac{m}{s^2}$ so, $\sqrt{\frac{kg\ m/s^2}{kg}m} = \sqrt{\frac{m^2}{s^2}} = \frac{m}{s}$.

    b. Since the speed is proportional to the square root of the height, it will take a quadrupling of the height to double the speed. So, the height needed to double the speed at contact with the ground is 4(0.46 m), or 1.84 m.

Example 2

At an instant of time a 2 kg block is sliding across a flat-horizontal surface at 4 m/s.

    a. If the block has a coefficient of kinetic friction of 0.15 with the surface, how far will it travel before coming to rest?

    b. How far will the same block travel across the same surface if the initial speed is doubled to 8 m/s?

Solution

    a. Step 1. Generate a simple sketch to define the two instants of time i and f, as shown in Figure 16.14.

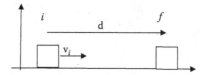

FIGURE 16.14   Example 2 sketch.

    Step 2. Apply conservation of energy:

$$K_i + U_i + W_{done} = K_f + U_f$$

The block slides along a horizontal surface, so the gravitational potential energies are set to zero, since $h_i = h_f = 0$ m. So, given that $U_i = U_f = 0$ J, they can both be cancelled, and conservation of energy results in $W_{done} = K_f - K_i$.

Since the work is done by friction and frictional force is in the opposite direction as the displacement $W_{done} = F_f\ d\cos(180° - 0°) = -\mu F_N\ d = -\mu\ m\ g\ d$ and $K_f - K_i = \frac{1}{2}\ mv_f^2 - \frac{1}{2}\ mv_i^2 = -\frac{1}{2}\ mv_i^2$, because $v_f = 0$ m/s. The two equations can be combined to give:

$$- \mu_k mgd = -\frac{1}{2}\ mv_i^2$$

This equation can be solved for d,

$$d = [\tfrac{1}{2}v_i^2]/[(\mu_k g)] = [\tfrac{1}{2}4\ m/s]^2/[(0.15)\ (9.8\ m/s^2)] = 5.44\ m$$

b. As can be seen in the equation for d in the line above, d is proportional to v^2.

Thus, a doubling of the velocity of the block will result in a quadrupling of the sliding distance. So, $(5.44)(4) = 21.76$ m

Example 3

A 30 kg skier enters the final section of a run traveling at 2.0 m/s. The final section of the run is 5.0 m long and has a vertical drop of 3.0 m. During the entire final section, the total (surface and air) force of friction on the skier is a constant 50 N.

    a. Find the velocity of the skier as she crosses the finish line at the end of the section.
    b. If all other parameters are kept constant, except that the length of the next run is increased to 10 m long, will the speed of the skier as she crosses the finish line be greater than, less than, or equal to her speed as she crosses the finish line in the original question?

Solution

    a. Step 1. Generate a diagram like the one given in Figure 16.15.

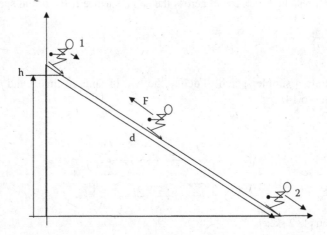

FIGURE 16.15   Example 3 sketch.

Step 2. Apply conservation of energy:

$$K_i + U_i + W_{done} = K_f + U_f$$

The initial kinetic energy is: $K_i = \frac{1}{2}mv_i^2 = \frac{1}{2}(30 \text{ kg})(2 \text{ m/s})^2 = 60$ J
  The initial potential energy is computed using the height (h), not the length of the run (d).

$$U_i = mgh_i = (30 \text{ kg})(9.8 \text{ N/kg})(3.0 \text{ m}) = 882 \text{ J}$$

Since the actual angle of the hill is not given, all we know is the force of friction is in the opposite direction as the displacement down the hill. Thus, the term with the $\cos(\theta_F - \theta_d) = \cos(180°)$, because the angle between the downhill and uphill directions is 180°.

$$W_{done} = F \, d \cos(\theta_F - \theta_d) = (50 \text{ N})(5 \text{ m}) \cos(180°) = -250 \text{ J}$$

The final potential energy is zero, because the height of the finish line is set at the zero height and the height of the beginning of this segment of the run is measured from the finish line height. $U_2 = 0J$

Conservation of energy gives:

$$K_i + U_i + W_{done} = K_f + U_f$$

$$60 \text{ J} + 882 \text{ J} + (-250 \text{ J}) = \tfrac{1}{2}(30 \text{ kg})v_f^2 + 0 \text{ J}$$

$$692 \text{ J} = \tfrac{1}{2}(30 \text{ kg})v_f^2$$

Solving gives the velocity at the finish line as:

$$v_2 = \sqrt{(692\,J)\left(\frac{1}{2}30\,kg\right)} = 6.79\frac{m}{s}$$

b. If the length of the hill is increased, the only term in the expression of the conservation of energy affected is the work due to friction. If the length of the run is doubled, the work due to friction is also doubled from −250 J to −500 J. Thus, the conservation of energy expression is

$$60 \text{ J} + 882 \text{ J} + (-500 \text{ J}) = \tfrac{1}{2}(30 \text{ kg})v_f^2 + 0 \text{ J}$$

Thus, the speed of the racer is decreased to v_f = 5.43 m/s

Example 4

A 2.0 kg block on a level surface is pressed against a spring, with a constant $k = 100$ N/m so that the spring is compressed 10 cm. The block is released, and the spring pushes it across the board. The coefficient of friction between the block and the surface is 0.20.

a. Find the distance the block travels across the surface as measured from the release point of the spring to the stopping point of the block.
b. If the spring is compressed twice as far, will the block slide twice as far?

Solution

a. Step 1. Generate a diagram like the one given in Figure 16.16.

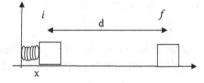

FIGURE 16.16 Example 4 sketch.

Step 2. Apply conservation of energy:

$$K_i + U_i + W_{done} = K_f + U_f$$

The initial and final kinetic energy is zero, because the block is at rest held against the compressed spring and at the end of the problem when it stops.

$$K_i = K_f = 0$$

The gravitational potential energy before and after is zero, because we can set the level surface as the zero height.

The spring is compressed $s = 10$ cm $= 0.1$ m, so the initial spring potential energy is: $U_i = \frac{1}{2} k s^2 = \frac{1}{2}$ $(100 \text{ N/m})(0.1 \text{m})^2 = 0.5$ J

The spring potential energy after is zero because the spring is no longer compressed.

Substituting into the conservation of energy gives:

$$K_i + U_i + W_{done} = K_f + U_f$$
$$0 \text{ J} + U_i + W_{done} = 0 \text{ J} + 0 \text{ J, so}$$
$$W_{done} = U_i = 0.5 \text{ J} = F_f d \cos(\theta)$$

and the force of friction is: $F_f = \mu N = \mu\, mg$, so 0.5 J $= \mu\, mg$ d $\cos(180°)$ solving for d gives a distance of 0.128 m

b. If the compression distance is doubled from 10 cm to 20 cm, the spring potential energy will quadruple, because the compression distance is squared in the potential energy expression,

$$U_i = \frac{1}{2} k d^2 = \frac{1}{2}(100 \text{ N/m})(0.2 \text{m})^2 = 2.0 \text{ J}$$

Or just multiply the original value of potential energy by 4 ($4 * 0.5$ J $= 2$ J). This leads to a quadrupling of the sliding distance to d $= 512$ cm, which is greater than $2 * 128$ cm $= 256$ cm.

16.7 POWER

Power is the time rate of change of the energy of a system. The average power associated with the amount of energy transferred ΔE_{01} during the time interval Δt_{01} is given in equation (16.10) as,

$$P_{avg,01} = \frac{\Delta E_{01}}{\Delta t_{01}} \tag{16.10}$$

Power is measured in units of energy per time, so it is measured in Joule per second, which is given the name the Watt (W).

$$1 \text{ W} = \frac{1\ J}{1\ s}$$

You may have seen this stamped on light bulbs and other electrical equipment. In the English system, the unit for power is also well known; it is horsepower (hp). The relationship between 1 hp and 1 W is:

$$1 \text{ hp} = 746 \text{ W.}$$

The transfer of energy is often done by some form of work, so in equation (16.10), the E_{01} is often found as the work done to transfer the energy, $\Delta E_{01} = W_{01}$. Since power is the time rate of the transfer of energy, often through doing work, it is an indication of how fast a person or machine can do a specific job or work. So, if there are two trucks, which can both do the same job, like tow a trailer up a hill, the more powerful truck can do the same work in a shorter time. The engine of the more powerful truck converts the potential energy in the gasoline to the kinetic energy of the trailer at a faster rate. A more powerful person would be able to stack a pile of boxes up on a shelf in a faster time than a less powerful person, but they both may be able to complete the work. The more powerful person does work to increase the potential energy of the boxes at a faster rate. Thus, power indicates how fast energy is converted from one type to another, often through work.

Example

A person of mass 62 kg climbs up a set of stairs in 3.00 seconds at a constant velocity, increasing her elevation by 3.6 m in that time. Find the average power being expended by the person in climbing the stairs, while she is climbing the stairs.

The person is converting chemical potential energy into gravitational potential energy. The total amount of energy converted into gravitational potential energy is her change in gravitational potential energy.

Step 1. Draw a before and after diagram, as given in Figure 16.17.

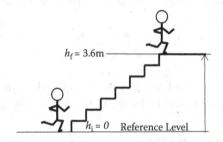

FIGURE 16.17   Power example sketch.

Remember that the initial before instant is the instant when she is at the bottom of the stairs and climbing, and the after instant is when she is at the top of the stairs where she still has the same velocity she had throughout the climb. The reference level is defined to be the elevation of the person's feet at time 0.

Step 2. Find the change in potential energy of the person for the climb:

$$\Delta U = U_f - U_i = mgh_1 - mgh_0 = mg(h_1 - h_0)$$
$$\Delta U = (62 \text{ kg})9.8 \text{ N/kg}(3.6 \text{ m} - 0 \text{ m})$$
$$\Delta U = 2187.16 \text{ J}$$

Step 3. Find the power by dividing the change in energy by the time it took to change that energy.

$$P = \frac{\Delta U}{\Delta t} = \frac{2187.16 \text{ J}}{3 \text{ s}} = 729.12 \text{ W}$$

16.8   ANSWER TO CHAPTER QUESTION

Starting with conservation of energy for the car skidding across the road:

$$K_i + U_i + W_{done} = K_f + U_f.$$

The potential energies are set to zero, since the road is level, so $h_i = h_f = 0$ m, so $U_i = U_f = 0$ J. So, the work done by friction is just the change in kinetic energy:

$$W_{done} = K_f - K_i$$

Plugging in the expressions for work due to friction and kinetic energy gives:

$$W_{done} = F_f d \cos(180°) = -F_f \, d = \frac{1}{2}mv_f^2 - \frac{1}{2}mv_i^2$$

Since the final velocity is zero: $-F_f \, d = -\frac{1}{2}mv_i^2$, the negatives cancel from each side and solving for the displacement gives:

$$d = \frac{1}{2}\frac{mv_i^2}{F_f}$$

Since the force of friction F_f is proportional to the normal force, F_N, by the expression $F_f = \mu F_N$, and for a car on a flat road the normal force has the same magnitude as the weight of the car ($F_g = mg$), the displacement of the car does not depend on the weight of the car in a skid.

$$d = \frac{1}{2}\frac{mv_i^2}{\mu mg} = \frac{v_i^2}{2\mu g}$$

Thus, the distance (d) is proportional to the velocity squared, so if the speed is doubled, the distance required to stop the car would be four (4x) times as much.

16.9   QUESTIONS AND PROBLEMS

16.9.1   MULTIPLE CHOICE QUESTIONS

The cart in Figure 16.18 has a mass of 2 kg, and all measurements of potential energy and height are made relative to point C, which is at the bottom of the track.

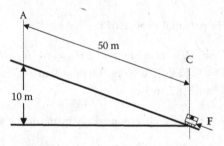

FIGURE 16.18   Multiple choice questions 1–3.

The track can be considered frictionless, and for questions 1–3 use the approximation of a gravitational field strength of $g = 10$ N/kg.

1. What is the gravitational potential energy of the cart at point A relative to point C?
   A.   100 J
   B.   200 J
   C.   500 J
   D.   1000 J
   E.   2000 J
   F.   5000 J

2. How much work is done on the cart by the person when the cart is moved from point C to point A?
   A. 100 J
   B. 200 J
   C. 500 J
   D. 1000 J
   E. 2000 J
   F. 5000 J

3. Assuming a constant applied force parallel to the incline, what is the magnitude of the force $\vec{F}$?
   A. 2 N
   B. 4 N
   C. 10 N
   D. 20 N
   E. 40 N
   F. 100 N

   In Figure 16.19, the cart has a mass of 2 kg and a kinetic energy of 20 J when it passes point A.

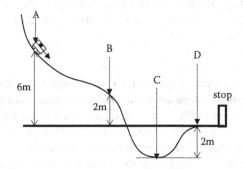

FIGURE 16.19 Multiple choice questions 4–6.

All measurements of potential energy and height are made relative to point D, which is the height of the track just before the stop. The track can be considered frictionless, and for **questions 4–7** use the approximation of a gravitational field strength of $g = 10$ N/kg.

4. What will be the kinetic energy of the cart at point B?
   A. 0 J
   B. 20 J
   C. 40 J
   D. 60 J
   E. 80 J
   F. 100 J
   G. 120 J
   H. 140 J
   I. 160 J
   J. 180 J

5. What will be the kinetic energy of the cart at point C?
   A. 0 J
   B. 20 J
   C. 40 J
   D. 60 J
   E. 80 J

   F. 100 J
   G. 120 J
   H. 140 J
   I. 160 J
   J. 180 J
6. What is the kinetic energy of the cart at point D?
   A. 0 J
   B. 20 J
   C. 40 J
   D. 60 J
   E. 80 J
   F. 100 J
   G. 120 J
   H. 140 J
   I. 160 J
   J. 180 J
7. A person on a roof throws a ball downward toward the ground below. The ball strikes the
   ground with 100 J of kinetic energy. The person throws an identical ball straight upward
   with the same initial speed. The ball leaves her hand at the same position it did in the first
   case. Neglecting air resistance, the second ball hits the ground with a kinetic energy of
   A. 100 J.
   B. less than 100 J.
   C. more than 100 J.
8. A bullet is fired with an initial speed of 100 m/s into a fixed block of wood, which is
   attached to a large table so it cannot move. The bullet goes 1 cm into the block before it
   stops. If the same type of bullet is shot out of a gun with a longer barrel, the bullet goes
   into the block of wood 4 cm before it comes to a stop. How fast was the bullet shot out of
   the longer barrel gun going before it entered the block? Assume that the force being
   exerted on the bullet by the wood while the bullet is both moving and in contact with the
   wood, is constant and is the same in both cases.
   A. 50 m/s
   B. 100 m/s
   C. 200 m/s
   D. 400 m/s
   E. 800 m/s
   F. 1600 m/s
9. A ball dropped from rest from a height h above ground level reaches the ground with 30
   J of kinetic energy. If air friction **was** significant as the ball fell, how much gravitational
   potential energy (using ground level as the reference level) did the ball have at its initial
   height h?
   A. more than 30 J
   B. less than 30 J
   C. exactly 30 J
10. Block 1 and block 2 are both at rest on a frictionless horizontal surface. The mass of
    block 2 is greater than the mass of block 1. A person pushes block 1 with a force of 2 N
    straight forward for 1 s. Starting at the same instant, another person pushes block 2 with
    a force of 2 N straight forward for that same 1 s time interval. At the end of that 1 s,
    which block, if either, has the greater kinetic energy?
    A. block 1
    B. block 2
    C. neither – both have the same amount of kinetic energy

16.9.2 Problems

1. Compute the work done by a force of 5 N applied to an object that is displaced 6 m along a flat, horizontal surface.
2. Compute the gravitational potential energy of a 2 kg object held 3 m above a point set as the zero-elevation reference point.
3. Compute the kinetic energy of a 4 kg object that is moving at a speed of 5 m/s.
4. A person pushes a cart a distance of 2 m, at constant velocity, up along a frictionless ramp that makes an angle of 16° with the horizontal. The person pushes the cart by means of a force of 14 N that is parallel to the ramp at an angle of 16° above the horizontal.
   a. Find the work done on the cart by the person.
   b. Find the mass of the cart.
5. A block with a mass of 500 g is placed on a level surface with which it has a coefficient of friction of 0.05 with no units. The block is then attached to a horizontal spring that has a force constant of 3.3 N/m and an unstretched length of 5.0 cm. The spring, while attached at one end to the block and the other end to a wall, is stretched to a length of 25 cm by pulling the block away from the wall. Then, with the block at rest, the person who pulled the block away from the wall lets go of the block. Compute the speed of the block at the point where the spring length is 10 cm.
6. A 95 kg skier starts from rest at a point 15 m above the horizontal section on a ski slope that ends in a long horizontal section. Due to some good snow conditions and ski waxing on the down slope, the friction is negligible. When the skier enters the horizontal section, he turns his skies to stop, and a constant force of 55 N acts on the skies in a direction directly opposite the motion of the skier and brings the skier to rest. How far does the skier travel on the horizontal section of the run? (Please see Figure 16.20.)

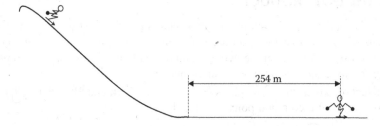

FIGURE 16.20 Problem 3.

7. A dart is loaded in a spring gun, with a spring having a force constant of 400 N/m. The dart compresses the spring 35 mm. The gun is pointed straight upward and fired. Neglect the force of the air on the dart. If the dart rises 2.2 m in the air above the point of release,
   a. How massive is the dart?
   b. How fast is the dart moving when it leaves the gun?
8. An 80 kg stuntman falls, from rest, off a building into an airbag, as shown in Figure 16.21.

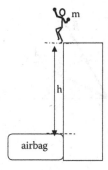

FIGURE 16.21 Problem 5.

The top of the airbag is exactly 10 m below the point from which the stuntman falls. If the force of air friction on the stuntman is an average value of 120 N on the stuntman during the fall, at what speed does he hit the top of the airbag?

9. A block with a mass of 1 kg is positioned at the top of a 1 m long ramp that is inclined so that the top of the ramp is 50 cm higher than the bottom of the ramp. Find the speed of the block at the bottom of the ramp if it is released from rest from the top of the ramp, if the ramp can be considered to be frictionless.

10. As depicted in Figure 16.22, a block with a mass of 1 kg is initially placed at point A and held against an uncompressing spring with a constant of 25 N/m. The block is then pushed to the left, compressing the spring a distance $s = 50$ cm and held at rest at point i.

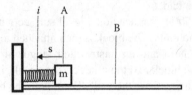

FIGURE 16.22 Problem 10.

The block is then released, so it is accelerated by the spring from point i to point A. After passing point A, the block moves at a constant speed across a horizontal frictionless surface toward point B. Find the speed of the block at the moment it passes by point B.

APPENDIX: THE DOT-PRODUCT

The dot product is another way of multiplying two vectors together that results in a magnitude that indicates how much the two vectors overlap. For example, the dot product of two vectors in the same direction is just the product of the two magnitudes, and the dot product of two perpendicular vectors is zero, because there is no overlap of components.

The dot product of two arbitrary vectors $\vec{A} = A_x\hat{i} + A_y\hat{j} + A_z\hat{k}$ and $\vec{B} = B_x\hat{i} + B_y\hat{j} + B_z\hat{k}$, with both their tails located at a common point is defined as:

$$\vec{A}\cdot\vec{B} = A_xB_x + A_yB_y + A_zB_z$$

where $\hat{i}$, $\hat{j}$, and $\hat{k}$ are gone and the result is just a number. The dot product is simply the sum of the products of the like components.

For vectors in the x-y plane, as depicted in Figure 16.23, the dot product is:

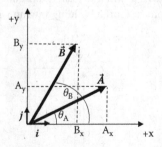

FIGURE 16.23 Dot product in 2-D.

$$\vec{\mathbf{A}} \cdot \vec{\mathbf{B}} = A_x B_x + A_y B_y$$

Which can be written:

$$\vec{\mathbf{A}} \cdot \vec{\mathbf{B}} = A\cos(\theta_A)B\cos(\theta_B) + A\sin(\theta_A)B\sin(\theta_B)$$

Given the following trigonometric identities:

$$\cos(A)\cos(B) = \frac{1}{2}[\cos(A + B) + \cos(A - B)]$$

$$\sin(A)\sin(B) = \frac{1}{2}[\cos(A - B) - \cos(A + B)]$$

The dot product of $\vec{\mathbf{A}}$ and $\vec{\mathbf{B}}$ results in:

$$\vec{\mathbf{A}} \cdot \vec{\mathbf{B}} = AB\frac{1}{2}[\cos(A + B) + \cos(A - B)] + AB\frac{1}{2}[\cos(A - B) - \cos(A + B)]$$

which can be rearranged to:

$$\vec{\mathbf{A}} \cdot \vec{\mathbf{B}} = AB\left\{ \frac{1}{2}[\cos(A + B)] + \frac{1}{2}[\cos(A - B)] + \frac{1}{2}[\cos(A - B)] - \frac{1}{2}[\cos(A + B)] \right\}$$

Notice that the cos(A + B) terms have the opposite sign in front of them and the cos(A − B) terms have the same sign in front, so the dot product for two vectors on the x-y plane is:

$$\vec{\mathbf{A}} \cdot \vec{\mathbf{B}} = AB[\cos(A - B)]$$

Given that work is defined as the dot product of the force, and the displacement for a typical situation like the one described in Figure 16.1 in which a crate is dragged across the floor by pulling on a rope that makes an angle θ with the horizontal, the work

$$W = \vec{F} \cdot \vec{d} = F_x d_x + F_y d_y$$

can be reduced to a simple equation. Since it is common to put the displacement along an axis, in this case the x-axis, the y-component of displacement is zero and x-component of the force is Fcos(θ). So, the expression for the work is:

$$W = \vec{F} \cdot \vec{d} = F\,d\,\cos(\theta)$$

where the angle is, $\theta = \theta_F - \theta_d$, the difference between the angle to the force and the angle to the displacement. In the diagram the force is at the angle θ and the angle of the displacement is $0°$, so the angle in the argument of the cosine is θ. This is a common situation, since the direction of the displacement is often along the x-axis or y-axis, so θ_d is often $0°$. If the displacement and the force are both at angles relative to the +x-axis, then just find the difference and plug it into the cosine. Therefore, the dot product provides a general expression for finding the work in three dimensions.

17 Momentum and Impulse

17.1 INTRODUCTION

Impulse and momentum are the next quantities at the center of another quantity-based analysis technique. The chapter begins with a force-based definition of impulse and the subsequent definition of momentum that evolves from an understanding of impulse. After some conceptual and numerical examples in which impulse and momentum are employed to understand some common events, the conservation of momentum is introduced. Through a series of examples, it is made clear that both conservation of momentum and conservation of energy are needed to analyze many physical systems, and both laws each have important roles in many applications.

17.2 CHAPTER QUESTION

When two objects collide, whether they are the same size or different sizes, like the truck and the car in Figure 17.1, and whether they are traveling at the same speed or different speeds, Newton's 3rd Law tells us that the force that each object exerts on the other is equal in magnitude and opposite in direction. So, what is different and what else, besides the force of one on the other, is the same? This question will be answered at the end of the chapter after the concepts of impulse and momentum are understood.

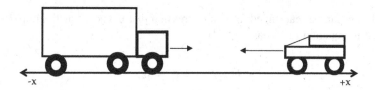

FIGURE 17.1 A collision between a truck and car.

17.3 IMPULSE

In the same way that the discussion of energy began with a definition of work as the product of force and displacement, the discussion of momentum starts with a definition of impulse, $\vec{J}$, as the product of the average net force, $\vec{F}_{Avg}$, and the time interval Δt_{01} during which the force is applied, as shown in equation (17.1) as

$$\vec{J} = \vec{F}_{Avg}\, \Delta t_{01} \tag{17.1}$$

A major difference between work and impulse is that work is a scalar and impulse is a vector. The definition of the vector $\vec{J}$ implies three scalar equations:

$$
\begin{aligned}
J_x &= F_{Avg,x}\, \Delta t_{01} \\
J_y &= F_{Avg,y}\, \Delta t_{01} \\
J_z &= F_{Avg,z}\, \Delta t_{01}
\end{aligned}
$$

We can typically define the x-axis to be parallel to the force and use just the first of these.

DOI: 10.1201/9781003308072-18

Newton's 2nd Law, $\vec{F}_{Net} = m\vec{a}$, applies at any instant in time. Over a time interval, from time t_0 to time t_1, Newton's 2nd Law tells us that $\vec{F}_{Avg} = m\vec{a}_{Avg}$, where $\vec{F}_{Avg}$ is the average net force applied to an object during that time interval Δt_{01} and $\vec{a}_{Avg}$ is the average acceleration of the object during the same time interval. Using the definition of average acceleration, which is the change in velocity during a time interval divided by the duration of that time interval, Newton's 2nd Law can be written as:

$$\vec{F}_{Avg} = m(\Delta\vec{v}/\Delta t_{01})$$

Multiplying both sides by Δt_{01} gives:

$$\vec{F}_{Avg}\,\Delta t_{01} = m\Delta\vec{v}$$

Rewriting $\Delta\vec{v}$ as $\vec{v}_1 - \vec{v}_0$ results in:

$$\vec{F}_{Avg}\,\Delta t_{01} = m(\vec{v}_1 - \vec{v}_0)$$
$$\vec{F}_{Avg}\,\Delta t_{01} = m\,\vec{v}_1 - m\,\vec{v}_0$$

Defining the vector quantity of momentum ($\vec{p}$) as the product of the mass, m, and velocity, $\vec{v}$, of an object in equation (17.2) as

$$\vec{p} = m\vec{v}, \tag{17.2}$$

Newton's 2nd Law can be rearranged into an expression that is known as the Impulse Momentum Relation, given in equation (17.3) as

$$\vec{F}_{Avg}\,\Delta t_{01} = \vec{p}_1 - \vec{p}_0 \tag{17.3}$$

Notice that both sides of equation (17.3) are ways to express the quantity of impulse, which can either be thought of as a change in momentum in equation (17.4) as

$$\vec{J} = \vec{p}_1 - \vec{p}_0, \tag{17.4}$$

or an average force applied over a time interval in equation (17.5) as

$$\vec{J} = \vec{F}_{Avg}\Delta t_{01}. \tag{17.5}$$

This Impulse Momentum Relation is similar to the relationship between work and kinetic energy; again, the major difference is that impulse and momentum are vectors, whereas work and kinetic energy are scalars. The direction of the momentum vector is the same as the direction of the velocity of the object, and the direction of the impulse is the same as the direction of the net force that results in the change in momentum. Momentum also differs from energy in that there are no different kinds of momentum – momentum cannot be transformed into another kind of momentum, it can only be transferred, and the only way to transfer it to a system consisting of a set of objects is by means of impulse. The vector relation $\vec{J} = \vec{p}_1 - \vec{p}_0$ has the following cartesian components:

$$J_x = p_{1x} - p_{0x}, \quad J_y = p_{1y} - p_{0y}, \quad J_z = p_{1z} - p_{0x}.$$

17.3.1 CONCEPTUAL EXAMPLES OF IMPULSE

The concept of impulse can be used to understand many of the ways people achieve the desired results in many situations. Following are a few examples of how impulse is applied.

1. In some cases, the right side of equation (17.3) is set by the circumstances, so to decrease the average applied force, the time of contact is increased.
   a. For example, air bags decrease the force felt by the driver or passenger in a car during an accident by controlling the duration of the time interval Δt_{01} that the person comes to rest. The change in momentum, and hence the impulse, is out of the control of those traveling in the car. The people in the vehicle have a mass as well as a velocity, and they will come to a stop abruptly during an accident. For a given impulse, if the duration of the time interval Δt_{01} is increased, the force is decreased. So, an air bag increases the duration of the impulse, which reduces the force.
   b. People bend their knees when they jump down from a height to the ground to increase the duration of the impulse similar to the way air bags work.
   c. When catching a ball in many sports such as baseball, lacrosse, or football, the players are taught to give with the ball when they catch it to increase the duration of the time interval Δt_{01}. You may have heard a coach telling you to catch with "soft-hands". This means to increase the time it takes to bring the ball to rest. Since the momentum change of the ball is the product of its mass and velocity just before you catch it to zero, when you catch it, increasing the duration of the time interval Δt_{01} decreases the force you must apply to the ball to bring it to rest.
2. In other situations, the goal is to create the largest change in momentum possible given a fixed average applied force, so the focus is to increase the duration of the time interval Δt_{01} in which a force is applied.
   a. In the case of hitting a baseball or golf ball, players are taught to follow-through with the hit. This achieves the greatest change in momentum because the applied force is applied for the greatest time, creating the largest change in momentum.
   b. The longer the barrel of an artillery cannon, the longer the range of the ordinance. Since the applied force due to the explosion is set by the caliber of the artillery, the longer the barrel, the more time the force due to the explosion can increase the speed of the ordinance. So, longer guns can send a projectile further.

17.3.2 NUMERICAL EXAMPLES OF IMPULSE

Example 1

Find the average force acting on a 1500 kg car traveling 20 m/s that hits a brick wall and comes to a stop in 0.2 s.

Solution

Step 1. Make a sketch of the situation, like the one in Figure 17.2, and define the coordinate system.

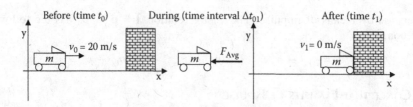

FIGURE 17.2   Impulse numerical example 1.

Step 2. Find expressions for the relevant components of the vectors.

$$v_{0x} = v_0 = 20 \text{ m/s} \quad v_{1x} = 0 \quad F_{Avg,x} = -F_{Avg}$$

Step 3. Employ the impulse momentum relation for the x components:

$$J_x = p_{1x} - p_{0x}$$

which can be written as:

$$
\begin{aligned}
F_{Avg,x}\, \Delta t_{01} &= p_{1x} - p_{0x} \\
F_{Avg,x} &= (p_{1x} - p_{0x})/\Delta t_{01} = (mv_{1x} - mv_{0x})/\Delta t_{01} = m\,(v_{1x} - v_{0x})/\Delta t_{01} \\
F_{Avg,x} &= 1500 \text{ kg } (0 \text{ m/s} - 20 \text{ m/s})/0.2 \text{ s} \\
F_{Avg,x} &= -150000 \text{ N}
\end{aligned}
$$

$\vec{F}_{Avg} = 150000$ N in the direction opposite that of the initial velocity of the car.

Example 2

A 0.1 kg baseball is thrown from the pitcher's mound so that it is moving horizontally at 20 m/s just before it comes in contact with the bat above home plate. The bat is in contact with the ball for .010 s. Upon losing contact with the bat, the ball's velocity is 30 m/s in the direction exactly opposite that of the velocity before it was hit by the bat; the ball is traveling horizontally, straight back at the pitcher. Compute the impulse delivered to the ball by the bat, and the average force of the bat on the ball while they are in contact.

Solution

Step 1. Make a sketch of the situation, as done in Figure 17.3, and define the coordinate system.

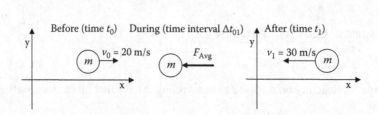

FIGURE 17.3   Impulse numerical example 2.

Step 2. Write the relevant components:

$$J_x = J_x \quad v_{0x} = v_0 = 20 \text{ m/s} \quad v_{1x} = -v_1 = -30 \text{ m/s}$$

Step 3. Use the impulse momentum relation (x-component only):

$$J_x = p_{1x} - p_{0x} = m \, v_{1x} - mv_{0x} = m(v_{1x} - v_{0x}) = .1 \text{ kg} \, (-30 \text{ m/s} - 20 \text{ m/s})$$
$$J_x = -5 \text{ kg·m/s} = -5 \text{ N·s}$$

$\vec{J} = 5$ N·s in the direction from home plate horizontally toward the pitcher
From the definition of impulse $\vec{J} = \vec{F}_{Avg} \, \Delta t_{01}$ we have:
$\vec{F}_{Avg} = \vec{J}/\Delta t_{01} = (5$ N·s in the direction from home plate horizontally toward the pitcher$)/(.01$ s$)$
$\vec{F}_{Avg} = 500$ N in the direction from home plate horizontally toward the pitcher.

17.3.3 IMPULSE AND NEWTON'S 3RD LAW

According to Newton's 3rd law, the two forces of one interaction are equal in magnitude and opposite in direction. This means that if object A is exerting a force on object B, then object B is exerting a force of the same magnitude but opposite direction of object A. Given that impulse is defined as shown in equation (17.5), this means that whenever object A delivers an impulse to object B, object B delivers an impulse of the same magnitude but opposite direction to object A. That is, $\vec{J}_{BA} = -\vec{J}_{AB}$).

Example

A truck traveling along a horizontal road in Figure 17.1 crashes into a car delivering an impulse of 5000 N·s to the car. Neglect the frictional force of the road on the vehicles during the collision.

 a. In that collision, what is the impulse delivered to the truck by the car?
 b. What is the change in the momentum of the car as a result of the collision?
 c. What is the change in the momentum of the truck as a result of the collision?

Solution

 a. By Newton's 3rd Law, the impulse delivered by the car to the truck $\vec{J}_{CT}$ is the negative of the impulse delivered by the truck to the car $\vec{J}_{TC}$.

$$\vec{J}_{CT} = -\vec{J}_{TC}$$
$$\vec{J}_{CT} = -5000 \text{ N·s}$$

 b. By the impulse momentum relation, the change in momentum $\Delta \vec{p}_{car}$ of the car is equal to the impulse delivered to the car:

$$\Delta \vec{p}_{car} = \vec{J}_{TC}$$
$$\Delta \vec{p}_{car} = 5000 \text{ N·s} = 5000 \text{ kg} \, \frac{m}{s}$$

 c. By the impulse momentum relation, the change in momentum $\Delta \vec{p}_{truck}$ of the truck is equal to the impulse delivered to the car:

$$\Delta \vec{p}_{truck} = \vec{J}_{CT}$$
$$\Delta \vec{p}_{truck} = -5000 \text{ N·s} = -5000 \text{ kg} \, \frac{m}{s}$$

17.4   CONSERVATION OF MOMENTUM

For a system of objects, the total momentum of the system is the vector sum of the individual momenta of the objects making up the system. When any two objects within the system interact with each other, there is only one time interval for the duration of the interaction, and the two forces of the interaction are equal and opposite, so the two impulses, each being the product of the average force acting on an object times the duration of the interaction, always add up to zero. This is Newton's 3rd Law of action reaction applied to a collision. Therefore, the momentum of the system never changes unless something from the surroundings delivers an impulse to one or more of the objects making up the system. For a system in which two objects interact, like in a collision, the net force is always zero, so by equation (17.5) $\left(\vec{\mathbf{J}} = \vec{\mathbf{F}}_{\text{Avg}} \Delta t_{01}\right)$ the net impulse is always zero. Therefore, for a collision the value of the impulse in equation (17.4) $\left(\vec{\mathbf{J}} = \vec{\mathbf{p}}_1 - \vec{\mathbf{p}}_0\right)$ is zero $(\vec{\mathbf{J}} = \mathbf{0})$, so equation (17.4) results in the expression of **conservation of momentum,** given in equation (17.6) as

$$\vec{\mathbf{p}}_1 = \vec{\mathbf{p}}_0. \tag{17.6}$$

The most common application of conservation of momentum involves the collision of two objects when, during the collision, no appreciable external impulses from the surroundings are delivered to either object. Then, the system is the two objects, and the law of conservation of momentum is simply,

$$\vec{\mathbf{p}}_{A0} + \vec{\mathbf{p}}_{B0} = \vec{\mathbf{p}}_{A1} + \vec{\mathbf{p}}_{B1}$$

which, in component form, can be written

$$
\begin{aligned}
p_{A0x} + p_{B0x} &= p_{A1x} + p_{B1x} \\
p_{A0y} + p_{B0y} &= p_{A1y} + p_{B1y} \\
p_{A0z} + p_{B0z} &= p_{A1z} + p_{B1z}
\end{aligned}
$$

In a case where all motion is along a line, like in a head-on collision, it is common to assign the x-axis to be parallel to the line of motion, and only the first of these equations is needed. In the case of a glancing blow, all the motion will be in the x-y plane, in which case only the first two of the x & y equations are needed. Clearly, collisions in three dimensions require all three equations, but in this text, only collisions in one and two dimensions will be considered.

17.4.1   Types of Collisions

For a specific generic one-dimensional collision, like the one described in Figure 17.4, the conservation of the x-components of the momentum vectors are:

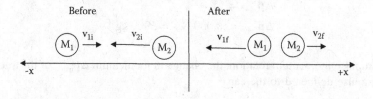

FIGURE 17.4   Generic 1-D collision diagram.

Notice that the force that ball 1 applies to ball 2 must be positive since ball 2 is moving in the negative direction before the collision and in the positive direction after the collision. Conversely, ball 1 is traveling in the positive direction before the collision, and after the collision it is moving in the negative direction, which would require the application of a force in the negative direction on ball 1.

In all collisions, momentum is conserved. In collisions that occur in two dimensions (2-D) or in three dimensions (3-D) momentum is conserved in each dimension independently. In all collisions, energy is conserved, but it is often difficult to keep track of the change in energy from one type to another during a collision. For example, in a collision between cars, the kinetic energy before the collision goes into deforming the vehicles, making the noise of the collision, and so on. This is why conservation of energy is not always the best method for investigating collisions. In some very special collisions, the kinetic energy before the collision is the same as the kinetic energy after the collision. This is not a common condition, but it can happen.

1. In an elastic collision, momentum and kinetic energy are both conserved. In these collisions, there is no deformation of the objects involved.
2. In an inelastic collision, only momentum is conserved. In these collisions, there is some deformation of the objects. The deformation may be temporary, like a collision between two rubber balls, or permanent, like a collision between two cars.

17.5   CONSERVATION OF MOMENTUM EXAMPLES

Example 1

A 1.0 kg piece of clay is thrown so that it collides at a speed of 3.0 m/s head on with a 2.0 kg piece of clay initially at rest. Find the velocity of the clay after impact if the pieces of clay stick together and travel as one object with a mass of 3.0 kg.

Solution

Step 1. Sketch a diagram like the one in Figure 17.5, which represents the situation before and after the collision. No coordinate system is provided in the problem statement, so the x direction is chosen for this example.

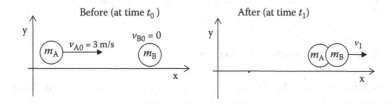

FIGURE 17.5   Conservation of momentum example 1.

Notice that this is an inelastic collision, since some of the mechanical energy is converted into thermal energy and energy of deformation during the collision.

Step 2. Write the components of the vectors.

$$v_{A0x} = v_{A0} = 3 \text{ m/s} \quad v_{B0x} = 0 \quad v_{1x} = v_1$$

Step 3. Write out the expression of conservation of momentum in the x direction for this collision.

$$p_{A0x} + p_{B0x} = p_{A1x} + p_{B1x}$$
$$m_A\, v_{A0x} + m_B\, v_{B0x} = m_A\, v_{1x} + m_B\, v_{1x}$$

Step 4. Solve for the unknown and plug in the values.

$$v_{1x} = \frac{m_A v_{A0x} + m_B v_{B0x}}{m_A + m_B}$$

$$v_{1x} = \frac{(1\ \text{kg})\, 3\frac{m}{s} + (2\ \text{kg})\, 0}{1\ \text{kg} + 2\ \text{kg}}$$

$$v_{1x} = 1\ \text{m/s}$$

Therefore, the final answer is that $\vec{v}_1 = 1$ m/s in the same direction as that of the original velocity of the 1 kg piece of clay.

Example 2

A 1.0 kg ball traveling forward at 1.00 m/s across a frictionless flat horizontal surface collides head on with a 2.0 kg ball initially at rest. After the collision, the 2.0 kg ball is observed to be moving forward at 0.600 m/s. Find the velocity of the 1.0 kg ball after the collision. Notice that these balls are not rolling; they are only traveling with translational, linear velocity. Rolling will be added in the next chapter.

Solution

Step 1. Sketch a diagram like the one in Figure 17.6, which represents the situation before and after the collision. No coordinate system is provided in the problem statement, so the x direction is chosen for this example.

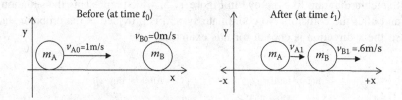

FIGURE 17.6   Conservation of momentum example 2.

Notice that the way ball A goes after the collision is not specified. Since it is less massive than ball B, it will probably bounce off and go the other way, but let's do the problem assuming that this is not known and see what happens.

Step 2. Write the components of the vectors.

$$v_{A0x} = v_{A0} = 1\ \text{m/s} \quad v_{B0x} = 0 \quad v_{B1x} = v_{B1} = .6\ \text{m/s}$$

Step 3. Apply conservation of momentum in the x direction gives:

$$p_{A0x} + p_{B0x} = p_{A1x} + p_{B1x}$$
$$m_A\, v_{A0x} + m_B\, v_{B0x} = m_A\, v_{A1x} + m_B v_{B1x}$$

Step 4. Solve for the unknown and plug in the values.

$$v_{A1x} = \frac{m_A v_{A0x} + m_B v_{B0x} - m_B v_{B1x}}{m_A}$$

$$v_{A1x} = \frac{(1\ kg)1\ \frac{m}{s} + (2\ kg)0 - (2\ kg)0.6\ \frac{m}{s}}{1\ kg}$$

$$v_{A1x} = -0.2\ m/s$$

Notice that the velocity of ball A is in the -x direction as we had suspected. If we had chosen this from the beginning, the answer calculated would be a positive value since the x-component of momentum of ball A after the collision would have had a negative sign, which could have, along with the 1.0 kg in the denominator, made the final answer positive. So, the lesson is do not double-count negatives. The final answer for the velocity of the 1 kg ball is: $\vec{v}_{A1}$ = .2 m/s backward.

Example 3

(An Elastic Collision Example) A 1.0 kg softball (m_1) traveling at 1.0 m/s across a flat horizontal surface collides head on with a 2.0 kg bowling ball (m_2) initially at rest. Find the velocity of each object after the collision if it is assumed that the collision is perfectly elastic. Please notice that these balls are not rolling.

Solution

Step 1. Sketch a diagram like the one in Figure 17.7, which represents the situation before and after the collision.

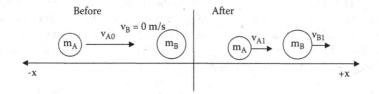

FIGURE 17.7 Conservation of momentum example 3.

Step 2. Write the components of the vectors.

$$v_{A0x} = v_{A0} = 1\ m/s \quad v_{B0x} = 0 \quad v_{A1x} = v_{A1} \quad v_{B1x} = v_{B1}.$$

Step 3. Apply conservation of momentum in the x direction, which gives:

$$P_{A0x} + P_{B0x} = P_{A1x} + P_{B1x}$$
$$m_A v_{A0x} + m_B v_{B0x} = m_A v_{A1x} + m_B v_{B1x}$$

Unfortunately, at this point the velocity of neither ball is known after the collision, so there are two unknowns, v_{A1x} and v_{B1x}. Therefore, another equation must be applied. Since this is a perfectly elastic collision, conservation of kinetic energy can be used to give:

$$\frac{1}{2}m_A v_{A0x}^2 + \frac{1}{2}m_B v_{B0x}^2 = \frac{1}{2}m_A v_{A1x}^2 + \frac{1}{2}m_B v_{B1x}^2$$

Solve the two equations for two unknowns.

Since $v_{B0x} = 0$, the momentum equation gives: $v_{B1x} = (m_A / m_B)(v_{A0x} - v_{A1x})$

Substituting this expression v_{B1x} into the kinetic energy equation gives:

$$\frac{1}{2}m_A\,v_{A0x}^2 + \frac{1}{2}m_B\,v_{B0x}^2 = \frac{1}{2}m_A\,v_{A1x}^2 + \frac{1}{2}m_B\left(\left(\frac{m_A}{m_B}\right)(v_{A0x} - v_{A1x})\right)^2$$

Multiplying through by 2, dividing by m_A, and factoring out the square gives:

$$v_{A0x}^2 + \frac{m_B}{m_A}v_{B0x}^2 = v_{A1x}^2 + \left(\frac{m_A}{m_B}\right)(v_{A0x}^2 - 2(v_{A1x})(v_{A0x}) + v_{A1x}^2)$$

Entering the numbers for $m_A = 1$ kg, $m_B = 2$ kg, $v_{A0x} = 1$ m/s and $v_{B0x} = 0$ m/s

$$1^2 + \frac{2}{1}\,0^2 = v_{A1x}^2 + \frac{1}{2}(1^2 - 2(v_{A1x})(1) + v_{A1x}^2)$$

Multiplying through by 2

$$2 + 0 = 2v_{A1x}^2 + 1 - 2v_{A1x} + v_{A1x}^2$$

and rearranging gives:

$$0 = -1 - 2v_{A1x} + 3v_{A1x}^2$$

Which is factorable to: $0 = (3v_{A1x} + 1)(v_{A1x} - 1)$, so the solutions to the problem are $v_{A1x} = -1/3$ m/s and $v_{A1x} = 1$ m/s. The second solution corresponds to the situation where the two objects miss each other, and the softball continues on with its speed unchanged. Notice that the direction of the softball after the collision was chosen in our diagram as the forward direction. The first solution is the one that is applicable to this collision. Substituting $v_{A1x} = -1/3$ m/s into the conservation of moment equation $v_{B1x} = (m_A/m_B)(v_{A0x} - v_{A1x}) = (1/2)(1 + 1/3)$ gives,

$$v_{B1x} = 2/3 \text{ m/s}$$

If you plug these numbers back into the conservation of momentum and the conservation of kinetic energy, they work. It is true that the algebra in this example is a bit complicated, but the idea is straightforward; both momentum and kinetic energy are conserved in perfectly elastic conditions. If there is a collision that is perfectly elastic, both conservation of momentum and conservation of kinetic energy can be applied to give two equations if needed. This is a special condition.

Example 4

(A 2-D Example) Ball A is moving at 2.0 m/s toward an initially at-rest ball B of the same mass. Both balls are not rolling, but are moving linearly, and the collision is not head on. After the collision, ball A is moving in a direction that makes an angle of 35° with its initial direction of travel, and ball B moves off at an angle of 50° from ball A's initial direction of travel. Find the speed of each ball at the instant after the collision.

Solution

Step 1. Draw a diagram similar to the one in Figure 17.8, in which a coordinate system is defined so that the initial direction of ball A is in the +x direction.

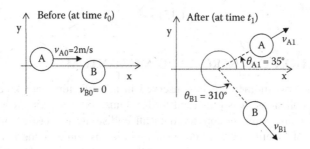

FIGURE 17.8 Conservation of momentum example 4.

Step 2. Write the components of the velocity vectors:

$$v_{A0x} = v_{A0} = 2 \text{ m/s} \qquad\qquad v_{A0y} = 0$$
$$v_{B0x} = 0 \qquad\qquad v_{B0y} = 0$$
$$v_{A1x} = v_{A1} \cos\theta_{A1} \qquad\qquad v_{A1y} = v_{A1} \sin\theta_{A1}$$
$$v_{B1x} = v_{B1} \cos\theta_{B1} \qquad\qquad v_{B1y} = v_{B1} \sin\theta_{B1}$$

Step 3. Apply conservation of momentum in the x and y directions separately:

$$p_{A0x} + p_{B0x} = p_{A1x} + p_{B1x} \qquad\qquad p_{A0y} + p_{B0y} = p_{A1y} + p_{B1y}$$
$$mv_{A0x} + mv_{B0x} = mv_{A1x} + mv_{B1x} \qquad\qquad mv_{A0y} + mv_{B0y} = mv_{A1y} + mv_{B1y}$$
$$v_{A0x} + v_{B0x} = v_{A1x} + v_{B1x} \qquad\qquad m0 + m0 = mv_{A1}\sin\theta_{A1} + mv_{B1}\sin\theta_{B1}$$
$$v_{A0} + 0 = v_{A1}\cos\theta_{A1} + v_{B1}\cos\theta_{B1} \qquad\qquad 0 = v_{A1}\sin\theta_{A1} + v_{B1}\sin\theta_{B1}$$
$$v_{A0} = v_{A1}\cos\theta_{A1} + v_{B1}\cos\theta_{B1} \qquad\qquad v_{B1} = -\frac{\sin\theta_{A1}}{\sin\theta_{B1}}v_{A1}$$

Substituting the equation for v_{B1} into the equation for v_{A0} yields:

$$v_{A0} = v_{A1}\cos\theta_{A1} + \left(-\frac{\sin\theta_{A1}}{\sin\theta_{B1}}v_{A1}\right)\cos\theta_{B1}$$

$$v_{A0} = v_{A1}\left(\cos\theta_{A1} - \frac{\sin\theta_{A1}}{\sin\theta_{B1}}\cos\theta_{B1}\right)$$

$$v_{A1} = \frac{v_{A0}}{\cos\theta_{A1} - \frac{\sin\theta_{A1}}{\sin\theta_{B1}}\cos\theta_{B1}}$$

$$v_{A1} = \frac{2\text{m/s}}{\cos 35\,° - \frac{\sin 35\,°}{\sin 310\,°}\cos 310\,°}$$

$$v_{A1} = 1.5379\,\frac{\text{m}}{\text{s}}$$

Substituting this result into the equation for v_{B1} yields:

$$v_{B1} = -\frac{\sin\theta_{A1}}{\sin\theta_{B1}}v_{A1}$$

$$v_{B1} = -\frac{\sin 35°}{\sin 310°}1.5379\,\frac{m}{s}$$

$$v_{B1} = 1.1515\,\frac{m}{s}$$

17.6 CONSERVATION OF ENERGY AND MOMENTUM

As mentioned previously, momentum is conserved in all collisions, but kinetic energy is only conserved in elastic collisions. It is important to distinguish between kinetic energy and the total energy, which, like momentum, is conserved in all collisions. In elastic collisions, the energy remains in the form of kinetic energy, but in most collisions some of the kinetic energy is converted into other types of energy. The energy is still conserved; it just changed forms to work or potential energy, as described concisely in the expression for conservation of energy:

$$K_i + U_i + W_{done} = K_f + U_f.$$

In some cases, conservation of energy is helpful, and in others the conversion from kinetic energy to another type of energy is too difficult to assess, and conservation of energy is not helpful. If you recall back to the previous chapter, conservation of energy is best applied to situations in which an object moves from one place to another. So, if a problem has information about how far or high objects move after a collision, then conservation of energy is often applicable. The following examples should help illustrate the types of questions in which conservation of energy should be used.

In Example 1 of the Conservation of Momentum section of this chapter, the velocity of two pieces of clay stuck together was found with conservation of momentum. This example asked: "A 1.0 kg piece of clay is thrown so that it collides at a speed of 3.0 m/s head on with a 2.0 kg piece of clay initially at rest. Find the velocity of the clay after impact if the pieces of clay stick together and travel as one object with a mass of 3.0 kg." This collision was perfectly inelastic, which means that the pieces of clay stick together and kinetic energy is not conserved.

The velocity of the two pieces of clay stuck together after the collision were found to be moving with a speed of 1 m/s. Given that there are no springs or elevation changes, the only kind of mechanical energy relevant to this problem is kinetic energy. Common sense tells us that the amount of mechanical energy converted to non-mechanical energy, such as thermal energy, is the initial kinetic energy of the two pieces of clay, which is the initial kinetic energy of the one that is moving, minus the final kinetic energy of the one object consisting of the two pieces of clay stuck together:

$$E_c = K_{A0} - K_{AB1}$$

$$E_c = \frac{1}{2}m_A v_{A0}^2 - \frac{1}{2}(m_A + m_B)v_1^2$$

$$E_c = \frac{1}{2}(1\text{kg})\left(3\frac{m}{s}\right)^2 - \frac{1}{2}(1\text{kg} + 2\text{kg})\left(1\frac{m}{s}\right)^2$$

$$E_c = 3\text{J}$$

In Example 2 of the Conservation of Momentum section of this chapter, the velocity of the 1 kg ball after the collision was found. The example stated, "A 1 kg ball traveling forward at 1m/s across a flat horizontal surface collides head on with a 2.0 kg ball initially at rest. After the collision, the 2.0 kg ball is observed to be moving forward at .600 m/s. Find the velocity of the 1.0 kg ball after the collision."

Solution

In this example the 1 kg ball, was labeled ball A, and the 2 kg ball, was labeled ball B. The solution was $\vec{v}_{A1} = -0.2$m/s backward, which means that the final speed of ball A is $v_{A1} = 0.2$m/s. All the mechanical energy is kinetic energy, and the amount converted to non-mechanical forms is the amount of mechanical energy that can be found. Before the collision, only ball A is moving, so only ball A has kinetic energy. Therefore, the energy after the collision is E_c.

$$E_c = K_{A0} - (K_{A1} + K_{B1})$$
$$E_c = \tfrac{1}{2}m_A v_{A0}^2 - \left(\tfrac{1}{2}m_A v_{A1}^2 + \tfrac{1}{2}m_B v_{B1}^2\right)$$
$$E_c = \tfrac{1}{2}(1\text{kg})\left(1\tfrac{m}{s}\right)^2 - \left(\tfrac{1}{2}(1\text{kg})\left(.2\tfrac{m}{s}\right)^2 + \tfrac{1}{2}(2\text{kg})\left(.6\tfrac{m}{s}\right)^2\right)$$
$$E_c = .12\text{J}$$

Note that $K_{A0} = \tfrac{1}{2}m_A v_{A0}^2 = \tfrac{1}{2}(1\text{kg})\left(1\tfrac{m}{s}\right)^2 = .5$J

$$\text{Percent Converted} = \frac{E_c}{K_{A0}} \times 100\% = \frac{.12\text{J}}{.5\text{J}} \times 100\%$$
$$\text{Percent Converted} = 24\%$$

Another class of problems involving energy and momentum consists of multipart problems in which, for one part of the process, it is appropriate to use conservation of momentum and for the other part it is appropriate to use conservation of energy. The best way to explain this is with an example of this kind of problem.

Example 1

A ballistic pendulum
A ballistic pendulum, sketched in Figure 17.9, is a device used to determine the speed of a bullet by shooting it horizontally into a pendulum bob. The bullet embeds itself in the bob. Then, the (bullet + bob) swing up to some maximum elevation.

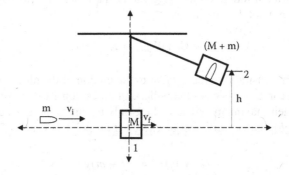

FIGURE 17.9 A ballistic pendulum.

Choosing the following instants in time:

t_0 = last instant before bullet hits bob
t_1 = first instant the bullet is fully lodged in bob, pendulum is still vertical but bob has forward velocity.
t_2 = instant (bullet + bob) reach maximum elevation

During the collision from time t_0 to time t_1, kinetic energy of the projectile, the bullet, is converted into forms of energy that are difficult to keep track of – in particular, into energy of deformation of the bob. Hence, it would be wrong to say that the mechanical energy before the collision is equal to the mechanical energy after the collision. However, there are no external impulses being delivered from the outside world to the bullet or the bob, although they certainly deliver impulses to each other, so it is correct to say that the total momentum of the projectile and the bob before the collision is equal to the total momentum of the projectile and the bob after the collision.

From t_1 to t_2, the earth is exerting a gravitational force on the (bullet + bob) that is not counteracted by any other force. As such, the earth delivers a significant impulse to the (bullet + bob). Thus, it would be wrong to say that the momentum of the (bullet + bob) at time t_2 is equal to the momentum of the (bullet + bob) at time t_1. However, neglecting frictional torque and air resistance, only the gravitational force of the earth does work on the (bullet + bob) from time t_1 to time t_2. Thus, the total mechanical energy at time t_1 is equal to the total mechanical energy at time t_2. Therefore, conservation of energy is employed from location 1 to location 2 to find the speed of the (bullet + bob) at time t_1. Conservation of momentum is used from time t_1 to time t_0 to find the speed of the projectile at time t_0.

Numerical Example

One way to find the speed of a bullet is to use a device called a ballistic pendulum. This device, depicted below in the diagram, is simply a large block of wood hung from the ceiling from a long and strong string. For this specific experiment, a bullet with a mass of 40 g = 0.04 kg is shot so that it has a total horizontal velocity as it strikes the wood block, which has a mass of 500 g = 0.500 kg. After the collision the bullet is lodged in the wood block, and the block and bullet rise a height h = 50 cm above the zero point. With this information, find the speed of the bullet just before it strikes the wood block.

Solution

Step 1. Make a sketch similar to Figure 17.9.
Step 2. Because of the information that is given in the problem, start with conservation of energy from point 1 to point 2:

$$K_1 + U_1 + W_{done} = K_2 + U_2$$

and since the tension is perpendicular to the motion of the block it cannot do any work, air friction can be neglected, the bullet and block come to an instantaneous stop when the reach their maximum height, and the only potential energy is gravitational potential energy (mgh), the conservation of energy from point 1 to point 2 becomes:

$$\frac{1}{2}(M + m)v_1^2 = (M + m)gh_2$$

The masses can be cancelled from both sides of the equation, and the velocity of the bullet and block (M + m) can be found:

$$v_1 = \sqrt{2gh_2} = \sqrt{2\left(9.8\frac{m}{s^2}\right)(0.5\ m)} = 3.13\frac{m}{s}$$

Step 3. Apply conservation of momentum for the collision between the bullet and the block.

$$mv_0 = (M + m)v_1$$

Plugging in the numbers gives the speed of the bullet before the collision as:

$$v_0 = \frac{(M + m)}{m}v_1 = \frac{(0.540)}{.04}3.13\frac{m}{s} = 42.26\frac{m}{s}$$

Steps 2 and 3 can be performed in a different order depending on the information given. If the speed of the bullet was given and the solution required the height of the (bullet + bob), the order of operations would be switched.

In this next example, the kinetic energy of the sliding skaters is converted to work due to friction. This example also illustrates the combination of 2-D conservation of momentum and conservation of energy.

Example 2

(Skaters Colliding): Two hockey players race at right angles to each other toward a puck, as described in the accompanying diagram. Player 1 has a mass of $m_1 = 150$ kg, and player 2 has a mass of $m_2 = 100$ kg. After colliding, the two players become tangled together and slide along the ice, tangled-up, at an angle of 35° from the original direction in which player 1 was traveling until they come to rest 2.0 m from the collision point on the ice. Assume the coefficient of friction between the sliding players and ice is 0.4, and calculate the speed of each player before the collision.

Solution

Step 1. Generate a diagram like Figure 17.10.

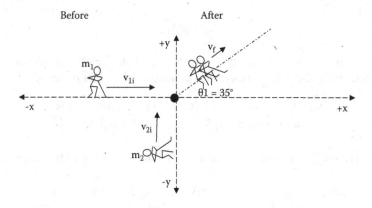

FIGURE 17.10   Colliding hockey players.

Step 2. Apply conservation of energy.

From the information given in the question, the speed of the players tangled together at the moment after the collision can be found with conservation of energy:

$$K_i + U_i + W_{done} = K_f + U_f$$

Since the ice should be level, the potential energy before and after the collision is zero along with the kinetic energy after the collision, so the conservation of energy reduces to:

$$K_i + W_{done} = 0.$$

Plugging in the equations for kinetic energy and work gives:

$$\tfrac{1}{2}(m_1 + m_2)v_f^2 + Fd\cos\left(\theta_f - \theta_d\right) = 0.$$

Since the force that stops the players is friction, it has the form:

$$F = \mu F_N = \mu(m_1 + m_2)g,$$

with a normal force with the same magnitude as the weight of the two hockey players. The expression from conservation of energy reduces to a solution for v_f:

$$v_f = \sqrt{\frac{-\mu(m_1 + m_2)gd\cos\left(\theta_f - \theta_d\right)}{\tfrac{1}{2}(m_1 + m_2)}}$$

Upon plugging in the known information and the angle of 180° for $(\theta_f - \theta_d)$ since the displacement and the force must be in opposite directions, the velocity of the players right after the collision is:

$$v_f = \sqrt{\frac{-0.4(150\ kg + 100\ kg)9.8\tfrac{N}{kg}2\ m\ cos(180°)}{\tfrac{1}{2}(150\ kg + 100\ kg)}}$$

which is

$$v_f = 3.96\ \text{m/s}$$

Step 3. Apply conservation of momentum to find the velocity of each player before they collide. First, the x & y components of final velocity can be found:

$$
\begin{aligned}
v_{fx} &= v_f \cos(35°) = 3.96\ \text{m/s} \cos(35°) = 3.24\ \text{m/s} \\
v_{fy} &= v_f \sin(35°) = 3.96\ \text{m/s} \sin(35°) = 2.27\ \text{m/s}
\end{aligned}
$$

Apply the conservation of momentum in the x and y directions separately, which gives:

$$
\begin{aligned}
x - \text{component:}\ & m_1 v_{1i} \cos(0°) + m_2 v_{2i} \cos(90°) = (m_1 + m_2)(3.24\ \text{m/s}) \\
y - \text{component:}\ & m_1 v_{1i} \sin(0°) + m_2 v_{2i} \sin(90°) = (m_1 + m_2)2.27\ \text{m/s}
\end{aligned}
$$

Notice that the angles of 0° and 90° indicate the direction the players are moving before the collision. Since the:

$$\sin(0°) = \cos(90°) = 0 \quad \& \quad \cos(0°) = \sin(90°) = 1$$

the x & y component equations of momentum reduce to:

$$\text{x} - \text{component: } m_1v_{1i} = (m_1 + m_2)3.24 \text{ m/s}$$
$$\text{y} - \text{component: } m_2v_{2i} = (m_1 + m_2)2.27 \text{ m/s.}$$

This gives final answers of: $v_{1i} = 5.4$ m/s and $v_{2i} = 5.7$ m/s.

17.7 ANSWER TO CHAPTER QUESTION

For the collision between the light car and a heavy truck, the forces are equal and opposite, and the total momentum of the system before the collision equals the total momentum after the collision. The important quantity that is different for the head-on collision in which the car and truck stick together and they both continue in the direction the truck was moving before the collision is the acceleration. Remember Newton's 2nd Law:

$$F_{net} = ma.$$

and that acceleration is a vector that indicates the rate of change in the velocity. Since the direction in which the car travels before and after the collision is opposite, as shown in Figure 17.11, the acceleration is larger since a negative is subtracted.

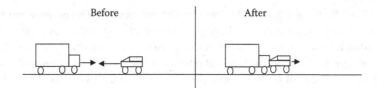

FIGURE 17.11 Colliding truck and car.

So, the starting point in understanding how the forces are the same is to recognize that the acceleration is very different.

17.8 QUESTIONS AND PROBLEMS

17.8.1 MULTIPLE CHOICE QUESTIONS

1. Block 1 and block 2 are both at rest on a frictionless horizontal surface. The mass of block 2 is greater than the mass of block 1. A person pushes block 1 with a force of 2 N straight forward for 1 s. Starting at the same instant, another person pushes block 2 with a force of 2 N straight forward for that same 1 s time interval. At the end of that 1 s, which block, if either, has the greater momentum?
   A. block 1
   B. block 2
   C. neither – both have the same momentum
2. A force of magnitude 4 N is applied to a 2 kg object to change its velocity from +2 m/s northward to +6 m/s northward. The force acts for a time interval of duration:

A. 1 s
B. 2 s
C. 4 s
D. 8 s

3. Object 1 is moving toward object 2, which is initially at rest and has the same mass as object 1. Is it possible for object 1 to hit object 2 with a glancing blow so that both objects only have a component of velocity after the collision, which is perpendicular to the direction of the original velocity of object 1 before the collision?

A. yes

B. no

4. Cart A and cart B are on a straight, horizontal, frictionless track. The mass of cart B is twice that of cart A. A person positions a horizontal spring of negligible mass between the two carts. It is not attached to either cart. Another person pushes the two carts together, compressing the spring, and holds them in that position while the first person distances her hand from the spring. The spring remains compressed between the two carts. Then, the second person releases both carts simultaneously from rest. The carts move apart from each other, and the spring drops to the track. How does the magnitude p_A of the momentum of cart A compare with the magnitude p_B of the momentum of cart B just after both carts have lost contact with the spring?

A. $p_A = \frac{1}{4} p_B$

B. $p_A = \frac{1}{2} p_B$

C. $p_A = p_B$

D. $p_A = 2 p_B$

E. $p_A = 4 p_B$

5. Cart A and cart B are on a straight, horizontal, frictionless track. The mass of cart B is twice that of cart A. A person positions a horizontal spring of negligible mass between the two carts. It is not attached to either cart. Another person pushes the two carts together, compressing the spring, and holds them in that position while the first person distances her hand from the spring. The spring remains compressed between the two carts. Then, the second person releases both carts simultaneously from rest. The carts move apart from each other, and the spring drops to the track. How does the speed v_A of cart A compare with the speed v_B of cart B just after both carts have lost contact with the spring?

A. $v_A = \frac{1}{4} v_B$

B. $v_A = \frac{1}{2} v_B$

C. $v_A = v_B$

D. $v_A = 2 v_B$

E. $v_A = 4 v_B$

6.–10. Multiple choice questions 6–10 are all based on the following description and the diagram in Figure 17.12.

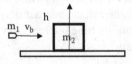

FIGURE 17.12 Multiple choice questions 6–10.

A small block of wood, with a mass of $m_2 = 1.6$ kg, is located on a flat surface, and the block is not attached to the surface. A bullet, with a mass of $m_1 = 30$ g, is fired directly at the block with a horizontal velocity of 400 m/s. The bullet blows directly through the block so that at the moment just after the collision the block and bullet are both moving directly to the right, with the bullet moving at 100 m/s and the block moving at a much lower speed than the bullet. Eventually, the block comes to rest due the frictional force between the block and the surface, and the bullet comes to rest due to air friction.

6. Is the magnitude of the momentum of the bullet the instant before the collision greater than, less than, or equal to the magnitude of the momentum of the bullet the instant after the collision?
   A. greater than
   B. less than
   C. equal to

7. Is the kinetic energy of the bullet the instant before the collision greater than, less than, or equal to the sum of the kinetic energy of the bullet and block the instant after the collision?
   A. greater than
   B. less than
   C. equal to

8. Is the magnitude of the momentum of the bullet the instant before the collision greater than, less than, or equal to the sum of the magnitudes of the momentum of the bullet and the block the instant after the collision?
   A. greater than
   B. less than
   C. equal to

9. If the block is switched out with one of the same mass, but made of a different material so that the bullet remains lodged in the block instead of passing through the block, will the momentum of the bullet after the collision be greater than, less than, or equal to the momentum of the bullet after the collision in the original configuration?
   A. greater than
   B. less than
   C. equal to

10. If the bullet in the original configuration was replaced with a rubber bullet, with the same mass and speed as the original bullet, so that it bounced off the block instead of penetrating through the block, will the distance the block will slide before coming to rest be greater than, less than, or equal to the distance the block slid in the original configuration?
   A. greater than
   B. less than
   C. equal to

17.8.2 PROBLEMS

1. A tennis ball of mass 58 g is traveling horizontally southward at 25 m/s when a tennis player hits it with her tennis racket. Just after she hits it, the ball is traveling 40 m/s horizontally northward. The racket is in contact with the ball for .0900 s.
   a. Find the impulse delivered to the ball by the racket.
   b. Find the average force of the racket on the ball.

2. Cart A of mass 1.00 kg and cart B of mass 2.00 kg are on one straight, horizontal, frictionless track. The two carts are moving toward each other. An x-y coordinate system has been defined such that the velocity of cart A before the two carts collide is

in the +x direction. The initial speed of cart B is 2.00 m/s. The carts collide. The carts are in contact with each other for .0500 s during the collision. After the collision, cart B is moving in the +x direction with a speed of 3.00 m/s.

   a. Find the impulse (magnitude and direction) delivered to cart A by cart B.
   b. Find the average force (magnitude and direction) being exerted on cart A by cart B during the collision.

3. A 500 g billiard ball moving at a speed of 2.2 m/s strikes an identical stationary 500 g billiard ball a glancing blow. After the collision, one ball is found to be moving at 1.1 m/s at an angle of 60° relative to the original line of motion. Find the velocity (speed and angle) of the other ball.

4. A ball of mass m traveling at 10 m/s makes a head-on elastic collision with a second ball. The second ball has mass $2m$ and is initially at rest. After the collision, the first ball recoils with a speed of 1.4 m/s in the direction opposite that of its initial velocity. What is the velocity of the second ball after the collision?

5. Two carts are placed on a horizontal track with negligible friction. Cart B is initially at rest and has a mass of 507 g. Cart A, which has a mass of 1016 g, is pushed forward toward cart B and released with a speed of 0.633 m/s. Assuming the collision is perfectly elastic, find the velocity of both carts after the collision.

6. A 30 g bullet is fired horizontally into a 1.2 kg wooden block that is initially at rest on a horizontal surface. The bullet becomes embedded in the block, and the block slides a distance of 7.5 m along the surface before stopping. If the coefficient of friction between block and surface is 0.66, what was the speed of the bullet before the collision?

7. A pendulum consists of a rubber ball of mass m_A attached to the end of a string, as shown in Figure 17.13.

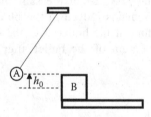

FIGURE 17.13 Problem 7.

The ball is hanging straight down, at its equilibrium position, and it is in contact with a block of mass m_B on a horizontal surface. The mass of the block is four times that of the ball ($m_B = 4m_A$). A person pulls the ball up and away from the block (see diagram) to a height h_0 of 20 cm above its equilibrium position. The rubber ball is released from rest. It swings back down and collides with the block. After the collision, the rubber ball bounces back to a maximum height of 6 cm above its equilibrium position, and the block slides a distance of 15 cm along the surface before coming to rest.

   a. Compute the velocity of the block the instant after the collision.
   b. Compute the percentage of the mechanical energy that is converted to non-mechanical energy during the collision.

8. Two hockey players race in directions at right angles to each other toward a puck. Player A has a mass of $m_A = 150$ kg, and player B has a mass of $m_B = 100$ kg. The players collide. Upon colliding, the two players become tangled together, and then they slide along the ice, tangled-up, in a direction that makes an angle of 35° with the original direction in which player A was traveling, until they come to rest 2.0 m from the collision point on the ice. Assume the coefficient of friction between the sliding players and the ice is 0.4, calculate the speed of each player at the last instant before the collision.

9. Cart A of mass 395 g is on a straight, frictionless, horizontal track and has a short, stiff, relaxed, ideal, massless, horizontal spring attached to one end. The force constant of the spring is 802 N/m. The relaxed length of the spring is 7.80 cm. While holding cart A in place, a person pushes cart B, which is on the same track and has a mass of 503 g, toward cart A, compressing the spring to a length of 4.50 cm, in between the two carts. The person releases the two carts from rest. The spring decompresses, pushing the two carts away from each other. How fast is each cart moving just after the unattached cart loses contact with the spring?

10. Particle A of mass 1.10 µg and charge 7.02 nC, and particle B of mass 2.58 µg and charge 9.82 nC, are released from rest at a separation of 1.00 mm. How fast is each particle going when the two particles are 5.00 cm from each other? No force other than the electrostatic force of the other particle, acts on either particle.

18 Angular Momentum and Rotational Kinetic Energy

18.1 INTRODUCTION

As in volume 1 of this textbook, the third introductory chapter of this volume is focused on the quantities employed to study the rotational counterparts of the linear quantities already introduced. In this volume, rotational impulse and angular momentum along with rotational kinetic energy are presented in this chapter. The chapter begins with a torque-based definition of rotational impulse and the subsequent definition of angular momentum. Conservation of angular momentum is developed from an understanding of rotational impulse, and several examples applying this conservation law are presented. Next, torque and rotational displacement are employed to develop an expression of rotational kinetic energy. The method used here is similar to that used to develop the concept of kinetic energy from force and linear displacement. Rotational kinetic energy is integrated into conservation of energy, and several examples are presented in which rotational quantities are used along with linear counterparts in the development of solutions.

18.2 CHAPTER QUESTION

Many figure skaters end their routines with a spin that starts with the skater spinning, then they draw their arms in tightly against their body, and they begin to spin much faster. The question is how is this possible if the skater does not push against the ice to spin faster? This question will be answered at the end of this chapter after the concept of angular momentum is developed and applied in the chapter.

18.3 ANGULAR IMPULSE

In the same way that the discussion of momentum began with a definition of impulse as the product of force and the duration of the time interval during which it acts, the discussion of angular momentum starts with a definition of angular impulse $\vec{J}_{Ang}$ given in equation (18.1) as the product of the average torque $\vec{\tau}_{Avg}$ and the time interval Δt_{01} during which the torque is applied,

$$\vec{J}_{Ang} = \vec{\tau}_{Avg}\Delta t_{01}. \tag{18.1}$$

The definition of the vector $\vec{J}_{Ang}$ (angular impulse) implies three vector components:

$$J_{Ang,x} = \tau_{Avg,x}\Delta t_{01} \quad J_{Ang,y} = \tau_{Avg,y}\Delta t_{01} \quad J_{Ang,z} = \tau_{Avg,z}\Delta t_{01}$$

The z-axis is normally defined to be parallel to the torque; in other words, the object is rotating in the x-y plane so the axis of rotation is along the z-axis. This is the orientation that will be employed for the entire chapter.

Newton's 2nd Law for rotational motion, $\sum \vec{\tau} = I\vec{\alpha}$, applies at any instant in time. Over a time interval, from time t_0 to time t_1, Newton's 2nd Law tells us that $\vec{\tau}_{Avg} = I\vec{\alpha}_{Avg}$, where $\vec{\tau}_{Avg}$ is the average net torque applied to an object during that time interval Δt_{01} and $\vec{\alpha}_{Avg}$ is the average angular acceleration of the object during the same time interval. Using the definition of average angular acceleration, the change in angular velocity during a time interval divided by the duration of that time interval, the expression of angular dynamics can be written as:

DOI: 10.1201/9781003308072-19

$$\vec{\tau}_{Avg} = I(\Delta\vec{\omega}/\Delta t_{01})$$

Multiplying both sides by Δt_{01} gives:

$$\vec{\tau}_{Avg}\Delta t_{01} = I\Delta\vec{\omega}$$

Rewriting the change in angular velocity $\Delta\vec{\omega}$ as $\vec{\omega}_1 - \vec{\omega}_0$

$$\vec{\tau}_{Avg}\Delta t_{01} = I(\vec{\omega}_1 - \vec{\omega}_0)$$
$$\vec{\tau}_{Avg}\Delta t_{01} = I\vec{\omega}_1 - I\vec{\omega}_0$$

Angular momentum $\vec{L}$ is defined as the product of the moment of inertia, I, and the angular velocity, $\vec{\omega}$, of an object so it can be expressed as shown in equation (18.2) as

$$\vec{L} = I\vec{\omega}. \tag{18.2}$$

Thus:

$$\vec{\tau}_{Avg}\Delta t_{01} = \vec{L}_1 - \vec{L}_0.$$

On the left, is the angular impulse $\vec{J}_{Ang}$, so the angular impulse and angular momentum relationship can be written as equation (18.3) as

$$\vec{J}_{Angular} = \vec{L}_1 - \vec{L}_0. \tag{18.3}$$

This relationship is similar to the relationship between linear impulse and linear momentum, given in equation (17.3). It applies to any rigid body, and since it is a vector relation, $\vec{J}_{Ang} = \vec{L}_1 - \vec{L}_0$. implies it is applied in its three vector components:

$$J_{Ang,x} = L_{1x} - L_{0x} \quad J_{Ang,y} = L_{1y} - L_{0y} \quad J_{Ang,z} = L_{1z} - L_{0z}$$

As with the skater, the z-axis is commonly defined to be parallel to the torque, and the z-component is the only one that will be used in this text.

Like linear momentum, angular momentum is a vector with its direction either clockwise or counterclockwise as observed from a point above the x-y plane and looking down from the +z-axis. In Figure 18.1, a disk is rotating counterclockwise as viewed from above the plane created by the x-y axes and along the +z-axis.

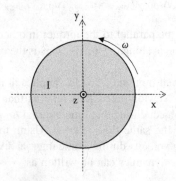

FIGURE 18.1   Rotating object.

This is defined as a positive rotation. If the disk was rotating clockwise as observed from above, this would be considered a negative rotation. This is just a standard labeling scheme, and clockwise and counterclockwise rotations are both equally significant ways for an object to rotate. A simple way to remember the signs of the rotation is the Right Hand Rule (RHR). As depicted in Figure 18.2, rotate the fingers of your right hand in the direction the object is rotating and the thumb of your right hand points in the direction of the angular momentum.

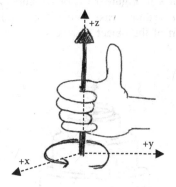

FIGURE 18.2 Right Hand Rule.

In Figure 18.2, an object rotating counterclockwise as observed from above the x-y plane is positive. This corresponds to the hand in Figure 18.2, in which the thumb points in the +z direction if the fingers of the right hand are rotating counterclockwise if the observer is looking down at the x-y plane from the +z axis. Another example of the Right Hand Rule and the direction of the angular momentum is the skater in Figure 18.3 is spinning in a counterclockwise direction as observed from above the skater, but it is easier to set the x-y plane on the surface of the ice, with the +x coming out at you and the +y to the right, and the angular momentum of the skater is in the +z direction.

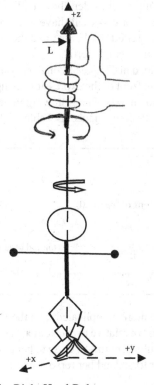

FIGURE 18.3 Rotating skater and the Right Hand Rule.

Remember that this is just a labeling, and there is nothing moving in the direction of your thumb; it is just an easy way to communicate which way the object is rotating.

18.4 CONSERVATION OF ANGULAR MOMENTUM

Consider the case of a single object that rotates about a single fixed axis. The object may be a single object, like the spinning disk in Figure 18.1, or a compound object, such as several disks stacked upon each other, like a wedding cake being decorated on a turntable as depicted in Figure 18.4, as long as every part of the object has the same angular velocity about one and the same axis.

FIGURE 18.4 Rotating wedding cakes.

In the compound object case, it is treated as a single object with the moment of inertia, with respect to the common rotational axis, as the sum of the moments of inertia of the objects making up the compound object. For example, consider the wedding cake spinning on the decorator turntable in Figure 18.4. Each of the four disks will have a moment of inertia equal to the product of one-half its mass times its radius squared, $I = \frac{1}{2} mr^2$, and the total moment of inertia of the entire cake will be the sum of all four moments of inertia.

The moment of Inertia (I) was defined in Chapter 13 on Rotational Dynamics. It has units of [kg m^2] and for different objects the fraction in front of the expression changes, since a different fraction of the mass is distributed around the rotational axis. The following table appeared in Chapter 13 for some simple objects and is included here for convenience as Table 18.1:

TABLE 18.1

Moments of inertia

Object	Moment of inertia (I)	Object	Moment of Inertia
Hoop	mr^2	Disk	$\frac{1}{2}mr^2$
Thin rod of length L about the center	$\frac{1}{12}mL^2$	Thin rod of length L about an end	$\frac{1}{3}mL^2$
Solid Sphere	$\frac{2}{5}mr^2$	Hollow Sphere	$\frac{2}{3}mr^2$

The compound object can also be more complicated, like the figure skater rotating in Figure 18.3. The skater consists of parts that can move relative to each other. If the skater is spinning about a vertical axis with arms extended, when she pulls her arms in close to her axis of rotation, as she brings them in, her arms are exerting a torque on her torso, and her torso is exerting torque on her arms. For a simple rotating object or a more complicated one, internal torques cancel each other out in pairs. Therefore, only external torques will change the angular momentum of the system, so the impulse momentum relation given in equation (18.3) for a single or a compound body is given by equation (18.3), where the torque in the expression is the net external torque. In situations where the net external angular impulse delivered to the system during some time interval, from t_0 to t_1, is zero, the angular momentum does not change, and the angular momentum is conserved, as expressed in equation (18.4) as

$$\vec{L}_0 = \vec{L}_1. \tag{18.4}$$

In this text, the z-axis will be chosen as the direction of the axis of rotation, so only the z-component of this equation will be applied in all examples and problems,

$$L_{0z} = L_{1z}.$$

18.4.1 Conservation of Angular Momentum Example

A dad, visiting at a local museum with his children, sits in a swivel chair, which is constructed with frictionless bearings. As part of the demonstration, he holds weights in his hands and holds his arms straight out from his sides. His children grab his outstretched arms and spin him so that he rotates at 2 revolutions per second. In this arrangement the system is the dad, the weights, and the part of the chair that rotates, and the moment of inertia, relative to the rotational axis, the vertical axis about which the chair spins, is 6 kg·m^2. When the dad brings the weights in close to his body, as specified in the directions of the exhibit, he spins at a rate of 3 revolutions per second. What must the new moment of inertia of the system be?

Solution

Step 1. Start by drawing a diagram, like the one in Figure 18.5.

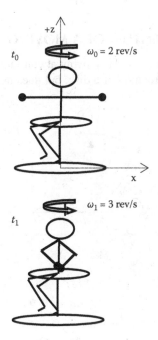

FIGURE 18.5 Example 1.

Be sure to define a coordinate system with the +z axis along the axis of rotation. The figure must depict the system both before and after the father pulls his arms in closer to his body. The sense of rotation was not specified, so in this figure it is specified as counterclockwise as viewed from above, so the angular moment is in the +z direction as given by the Right Hand Rule.

Step 2. Find the z-components of the angular velocities. Since the magnitudes of the angular velocities are given in rev/s, the first step is to convert them to rad/s. Since each revolution is 2π rad, the conversion is simple, multiplication of 2π to give:

$$\omega_0 = 2\frac{\text{rev}}{s} \cdot \frac{2\pi\ \text{rad}}{\text{rev}} = 4\pi\frac{\text{rad}}{s}$$
$$\omega_1 = 3\frac{\text{rev}}{s} \cdot \frac{2\pi\ \text{rad}}{\text{rev}} = 6\pi\frac{\text{rad}}{s}$$

The π is left in the expressions to avoid unnecessary rounding errors.

With the choice of the direction of rotation along the +z-axis, by the Right Hand Rule, the z-components of the angular velocities are:

$$\omega_{0z} = \omega_0 = 4\pi\frac{\text{rad}}{s}$$
$$\omega_{1z} = \omega_1 = 6\pi\frac{\text{rad}}{s}$$

Step 3. Apply conservation of angular momentum for the z-components and solve for the unknowns.

$$L_{0z} = L_{1z}$$
$$I_0\omega_{0z} = I_1\omega_{1z}$$
$$I_1 = \frac{\omega_{0z}}{\omega_{1z}}I_0$$
$$I_1 = \left(\frac{4\pi\ \text{rad}/s}{6\pi\ \text{rad}/s}\right)6\frac{\text{kg}}{\text{m}^2}$$
$$I_1 = 4\frac{\text{kg}}{\text{m}^2}$$

18.5   THE ANGULAR MOMENTUM OF A MOVING PARTICLE

Consider a particle, with a mass of m, that is moving at constant velocity, v, along a straight path that is perpendicular to an axis of rotation of a disk, but does not intersect the axis of rotation, as depicted in Figure 18.6.

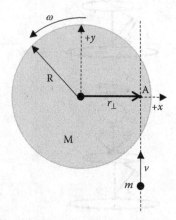

FIGURE 18.6   Angular momentum of an object moving with a linear velocity.

The coordinate system has an origin on the axis of rotation. The +x direction is to the right, and the +y axis is toward the top of the page, as shown in Figure 18.6. With this orientation, the +z-axis is out of the page, as given by the Right Hand Rule. If the disk is mounted on a frictionless axle and allowed to rotate freely about its axis and the particle collides with the disk at point A, the disk will rotate. The moment of inertia of the system after the collision, is a combination of the moment of inertia of the disk, $\frac{1}{2}MR^2$, and the moment of inertia of a particle, $m\,r_\perp^2$, at a rotational distance of $r_\perp$. So, if the disk and particle are spinning after the collision, at t_1, with an angular velocity of ω, the angular momentum of the system after the collision is:

$$L_{1,z} = \left(\frac{1}{2}MR^2 + m\,r_\perp^2\right)\omega_1.$$

Thus, by conservation of angular momentum, the system must have had an angular momentum before the collision. That angular momentum is associated with the velocity of the particle relative to the axis of rotation. As such, an angular momentum is assigned to the particle with a magnitude of $L_0 = I\,\omega_0$. Since the angular velocity is related to the linear velocity by $v = r_\perp\omega$, and the moment of inertia of the particle is $mr_\perp^2$ when it hits the disk and the disk begins to spin, the z-component of the angular momentum of the particle relative to the axis of rotation is

$$L_{1,z} = (mr_\perp^2)\omega_{1,z} = (mr_\perp^2)\left(\frac{v_{1,z}}{r_\perp}\right) = m\,r_\perp v_{1,z} = r_\perp p_{1,z}$$

So, the z-component of the angular momentum of a particle moving with a linear momentum $p_{1,z}$ at a relative perpendicular distance of $r_\perp$ from the axis of rotation of an object is given by equation (18.5) as

$$L_{1,z} = r_\perp p_1. \tag{18.5}$$

The product $r_\perp\,p$ is the magnitude of the cross product $\vec{r} \times \vec{p}$, and, that cross product also yields the correct direction for the angular momentum. So, equation (18.5) can be written in a more general format using the cross product as given in equation (18.6) as

$$\vec{L} = \vec{r} \times \vec{p}. \tag{18.6}$$

Example

As described in Figure 18.7, a bullet of mass 20.0 grams is traveling horizontally at 1005 m/s in the direction 31.5 degrees, directly toward a point on the outer edge of a wooden horizontal disk of radius .650 m that is mounted so that it is free to rotate about a fixed vertical axis through the center of the disk.

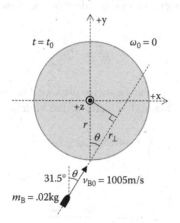

FIGURE 18.7 Example of angular momentum of an object moving with a linear velocity at $t = t_0$.

The disk is initially at rest. The bullet embeds itself in the disk, and after the collision, the disk with the bullet embedded is rotating about the axis with a constant angular velocity. The moment of inertia, with respect to the axis of rotation of the disk with the bullet embedded in the disk, is 4.25 kg·m^2. How fast is the disk with the bullet embedded in it rotating, in rad/s, after the collision?

Solution

Step 1. Given that the before diagram, at $t = t_0$, is given with the problem, the after diagram at $t = t_1$ is sketched in Figure 18.8.

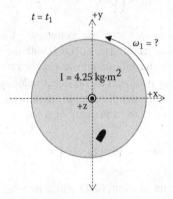

FIGURE 18.8   Example of angular momentum of an object moving with a linear velocity at $t = t_1$.

Notice in both diagrams there is a circle with a dot in it at the center of the disk and +z next to the circle and dot. This symbol indicates an axis coming out of the page that is the +z direction. This is meant to represent the point of an arrow coming out of the page.

Step 2. Write out the z-components of the angular momenta for the problem.

$$L_{0z} = r_\perp p_{B0}$$
$$L_{1z} = I\omega_1$$

Step 3. Apply conservation of angular momentum in the z direction and solve for ω_1. Because, by the Right Hand Rule, both $\mathbf{L}_0$ and $\mathbf{L}_1$ are in the +z direction, conservation of angular momentum in the z direction is simply:

$$L_0 = L_1$$
$$r_\perp p_{B0} = I\omega_1$$

Notice that the $r_\perp$ is equal to $r \sin\theta$ by the geometry presented in Figure 18.7 and momentum of the bullet is just the product of the mass and velocity of the bullet, so the expression becomes,

$$r \sin\theta m_B v_{B0} = I\omega_1$$

Solving for ω_1 gives:

$$\omega_1 = \frac{r \sin\theta m_B v_{B0}}{I}.$$

Plugging in the numbers to find the angular velocity, ω_1, of the system after the collision gives

$$\omega_1 = \frac{.65\text{m}\,\sin(31.5°).02\,\text{kg}(1005\text{m/s})}{4.25\text{kg}\cdot\text{m}^2} = 1.6062\frac{\text{rad}}{\text{s}}.$$

18.6 ROTATIONAL KINETIC ENERGY

A body in pure rotation with an angular velocity $\vec{\omega}$ consists of a set of particles, each of which is moving in a circle. Assume that an object is made up of N particles, numbered from 1 through N. It is common to use the subscript i to represent the ith particle in the object. The kinetic energy, K, of the rotating object is the sum of the kinetic energies of all the particles that make up the object, which can be expressed as

$$K = \sum_{i=1}^{N} \frac{1}{2}m_i v_i^2 = \frac{1}{2}m_1 v_1^2 + \frac{1}{2}m_2 v_2^2 + \frac{1}{2}m_3 v_3^2 + \cdots + \frac{1}{2}m_N v_N^2$$

The ith particle is a distance r_i out from the axis of rotation, meaning it is moving on a circle of radius r_i and has a speed of $v_i = r_i\omega$. Note that there is no subscript on the angular velocity term ω because every particle making up the object has the same angular velocity. Substituting $v_i = r_i\omega$ into our expression for the kinetic energy yields:

$$K = \sum_{i=1}^{N} \frac{1}{2}m_i (r_i\omega)^2$$

which can be written

$$K = \frac{1}{2}\left(\sum_{i=1}^{N} m_i r_i^2\right)\omega^2$$

The quantity in parentheses is, by definition, the moment of inertia I of the object. Thus, the kinetic energy of an object in pure rotation can be expressed in equation (18.7) as

$$K = \frac{1}{2}I\omega^2. \tag{18.7}$$

For the case of an object that is moving through space with a translational linear velocity, v, and spinning with an angular velocity, ω, at the same time, the total kinetic energy of the object is the sum of the translational kinetic energy, K_T, and the rotational kinetic energy, K_R is given in equation (18.8) as

$$K = K_T + K_R = \frac{1}{2}mv^2 + \frac{1}{2}I\omega^2 \tag{18.8}$$

where the mass, m, is the mass of the entire object and the speed, v, is the linear speed with which the axis of rotation is moving through space. When the center of mass of the object lies on the axis of rotation, the speed v is also the speed with which the center of mass is moving through space.

18.6.1　Examples of Rotational Kinetic Energy

Example 1

Recall the conservation of angular momentum example of the dad on the swivel chair at the museum, in which the system consists of the dad plus the weights in his hand plus the part of the chair that was rotating had a moment of inertia of 6 kg·m^2 and was spinning at 2 rad/s. He pulled his arms in, and the magnitude of his angular velocity changed to 3 rad/s. Using conservation of angular momentum, we determined that the new moment of inertia of the system was 4 kg·m^2. *Now* the question is, what was the change in the kinetic energy (if any) of the system?

Solution

The change in anything is the final value minus the initial value so, using the notation ΔK_{01} for the change in the kinetic energy from time 0 to time 1, we have:

$$\Delta K_{01} = K_1 - K_0$$

where:

$$K_1 = \frac{1}{2}I_1\omega_1^2 = \frac{1}{2}(4 \text{ kg·m}^2)\left(3\frac{\text{rad}}{\text{s}}\right)^2 = 18 \text{ J}$$

and

$$K_0 = \frac{1}{2}I_0\omega_0^2 = \frac{1}{2}(6 \text{ kg·m}^2)\left(2\frac{\text{rad}}{\text{s}}\right)^2 = 12 \text{ J}$$

so:

$$\Delta K_{01} = 18 \text{ J} - 12 \text{ J}$$
$$\Delta K_{01} = 6 \text{ J}$$

This represents a 50% increase in the system's kinetic energy, and one might well ask where that energy came from. In using his muscles to pull the weights in, the dad converted chemical potential energy to mechanical energy. Note that it is wrong to assume that the final kinetic energy is the same as the initial kinetic energy in solving the original problem.

Example 2

The situation shown in Figure 18.6 is associated with an example in which a bullet is shot at a disk at rest and the disk and bullet together begin to spin.

Solution

Referring back to Figures 18.7 and 18.8, the total kinetic energy of the system before the collision is just the translational kinetic energy of the bullet before the collision:

$$K_0 = \frac{1}{2}mv_{B0}^2 = \frac{1}{2}(.02 \text{ kg})(1005 \text{ m/s})^2$$
$$K_0 = 10100.25 \text{ J}$$

The total kinetic energy of the system after the collision is just the rotational kinetic energy of the disk plus bullet after the collision.

$$K_1 = \tfrac{1}{2}I\omega_1^2 = \tfrac{1}{2}(4.25 \text{ kg}\cdot\text{m}^2)\left(1.6062\tfrac{\text{rad}}{s}\right)^2$$

$$K_1 = 5.4822 \text{ J}$$

The amount of energy converted into other forms, mostly thermal energy, is:

$$E_c = K_1 - K_0 = 10100.25 \text{ J} - 5.4822\text{J} = 1.0094.77 \text{ J}$$

Note that in this case, almost all (over 99.9%) of the kinetic energy is converted into other forms of energy.

18.6.2 KINETIC ENERGY OF A ROLLING SYMMETRICAL OBJECT

A rolling symmetrical round object whose center of mass lies on its axis of rotation, such as a ball, hoop, or a disk, is both moving through space and spinning. Therefore, equation (18.8) represents the kinetic energy of the system. If the object is rolling without slipping, there is a simple relationship between how fast it is moving forward and how fast it is spinning. If an object is rolling without slipping, then the distance, d, it travels in one rotation at a constant angular velocity, w, is its circumference $2pr$. If the time for one complete rotation is just the period, T, the speed of the object is given by

$$v = \frac{d}{\Delta t} = \frac{2\pi r}{T}$$

Since the magnitude of angular velocity is $\omega = \frac{1 \text{ rev}}{T} = \left(\frac{1 \text{ rev}}{T}\right)\left(\frac{2\pi \text{ rad}}{1 \text{ rev}}\right) = \frac{2\pi \text{ rad}}{T}$, the period of a rotation is $T = \frac{2\pi \text{ rad}}{\omega}$. Substituting this into our expression for v results in $v = \frac{2\pi r}{\left(\frac{2\pi \text{ rad}}{\omega}\right)}$ $v = \frac{2\pi r}{\left(\frac{2\pi \text{ rad}}{\omega}\right)}$, which reduces to the simple expression of equation (18.9),

$$v = r\omega \tag{18.9}$$

Thus, the faster a rolling object is spinning, the faster it is moving forward.

Example

A uniform solid marble with a mass of 10 g and a radius of 1 cm, starting at rest, rolls without slipping, down a 1 m long inclined plane that is tilted at an angle of 30° to the horizontal. Compute the speed of the marble at that instant when it reaches the bottom of the inclined plane.

Solution

Step 1. Draw a before and after diagram, as shown in Figure 18.9.

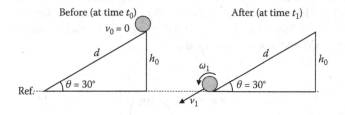

FIGURE 18.9 Rolling example.

Step 2. Apply conservation of mechanical energy:

$$K_{T0} + K_{R0} + U_0 = K_{T1} + K_{R1} + U_1$$
$$0 + 0 + mgh_0 = \tfrac{1}{2}mv_1^2 + \tfrac{1}{2}I\omega_1^2 + 0$$
$$mgh_0 = \tfrac{1}{2}mv_1^2 + \tfrac{1}{2}I\omega_1^2$$

Apply equation (18.9), for an object that is rolling without slipping, to get $\omega_1 = v_1/r$ so the expression of conservation of energy becomes:

$$mgh_0 = \frac{1}{2}mv_1^2 + \frac{1}{2}I\left(\frac{v_1}{r}\right)^2$$

Also, from Table 18.1, for a uniform solid sphere, the moment of inertia is $I = \tfrac{2}{5}mr^2$, so:

$$mgh_0 = \tfrac{1}{2}mv_1^2 + \tfrac{1}{2}\left(\tfrac{2}{5}mr^2\right)\left(\tfrac{v_1}{r}\right)^2$$
$$gh_0 = \tfrac{1}{2}v_1^2 + \tfrac{1}{5}v_1^2$$
$$gh_0 = \tfrac{7}{10}v_1^2$$
$$v_1 = \sqrt{\tfrac{10}{7}gh_0}$$

From the geometry of the configuration, we know that $h_0 = d \sin \theta$. Thus:

$$v_1 = \sqrt{\frac{10}{7}gd \sin \theta}$$

Step 3. Plug in the known values and evaluate:

$$v_1 = \sqrt{\tfrac{10}{7}\left(9.8\tfrac{N}{kg}\right)(1m)\sin 30°}$$
$$v_1 = 2.64575\tfrac{m}{s}$$

Note that the answer did not depend on the radius or mass of the marble. Any size uniform sphere would have the same speed at the bottom as long as it rolls down the inclined plane without slipping.

Also note that if the surface of the inclined plane were frictionless, instead of rolling without slipping, the marble would slide down the ramp without rotating. At the bottom, all the energy would be translational kinetic energy, and given that it came down from the same height, the total energy would have to be the same as in the rolling without slipping case. Hence, the marble would have to have more translational kinetic energy at the bottom than it does in the rolling without slipping case.

For comparison, the following is the calculation of an object sliding without friction down the same inclined plane. Conservation of energy gives:

$$K_{T0} + U_0 = K_{T1} + U_1$$
$$0 + mgh_0 = \tfrac{1}{2}mv_1^2 + 0$$
$$mgh_0 = \tfrac{1}{2}mv_1^2$$

The *m*s will cancel, and the speed at the end of the inclined plane is:

$$v_1 = \sqrt{2gh_0}$$

and the height $h = 0.5$ m results in a speed at the bottom of the ramp of $v_1 = 3.13$ *m/s*.

So, rolling slows the marble down as compared to an object moving without friction. This makes sense, because friction causes the marble to roll in the first place. The friction on the bottom of the marble rotates the marble and converts some of the gravitational potential energy to rotational kinetic energy, so the linear kinetic energy must be less than the object sliding without friction.

18.7   ANSWER TO THE CHAPTER QUESTION

Conservation of the angular momentum provides the conceptual framework needed to find the change in the angular velocity of the skater as she brings in her hands closer into her body. As she brings her hands and the rest of her body closer to the line of rotation, she decreases her moment of inertia; thus, her angular velocity increases to keep angular momentum constant (Figure 18.10).

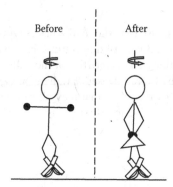

FIGURE 18.10   The spinning ice skater of the chapter question.

Given that the magnitude of angular momentum is $L = I\,\omega$, for the skater that we used as an example: By pulling her arms in, she decreases her moment of inertia, I. If the ice exerts no torque on her while she is doing this, for the product $I\,\omega$ to remain the same, ω must increase; in other words, she must spin faster. If you have ever seen a figure skater do this, you know that when she pulls her arms, in she does indeed spin faster.

18.8   QUESTIONS AND PROBLEMS

18.8.1   MULTIPLE CHOICE QUESTIONS

1. An ice skater spinning about a vertical axis with arms extended outward pulls her arms in close to her torso. What happens to the skater's angular momentum about her axis of rotation as she pulls her arms in? (Assume that during the process, the ice exerts no torque on the skater.)
   A. It increases.
   B. It decreases.
   C. It stays the same as what it was.

2. If the net torque on a rotating object is zero, will that object's angular momentum necessarily be constant in both magnitude and direction?
   A. yes
   B. no

3. The rotating remnants of a star that has undergone a supernova explosion collapse to become a neutron star with a much smaller size than the remnants, but the same mass. What happens to the angular velocity of the system consisting of all the remnants of the star from before to after the collapse?
   A. It increases.
   B. It decreases.
   C. It stays the same as what it was.

4. Three golf balls are manufactured with the same mass and radius (Figure 18.11).

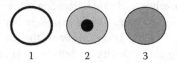

FIGURE 18.11 Multiple choice question 4.

Ball 1 is a hollow shell of a dense material, ball 2 has a light outer core wrapped around a very dense inner core, and ball 3 is a solid uniform sphere made out of a material with an intermediate density. The rolling characteristics of the balls are tested by letting them roll down an incline. The three balls are released simultaneously, side-by-side from a position near the top of the incline. If we assume that they all roll without slipping, in what order do they reach the bottom of the incline (first to last).
   A. 1, 2, 3
   B. 2, 3, 1
   C. 2, 1, 3
   D. 3, 1, 2
   E. 3, 2, 1
   F. all at the same time

5. During a demonstration done in a physics class, depicted in Figure 18.12, a student sits in a swivel chair with arms and legs extended.

FIGURE 18.12 Multiple choice question 5.

Another student pushes the sitting student so that the sitting student begins spinning about a vertical axis. The sitting student then brings her legs and arms in, after which, she is spinning faster. During this demonstration, neglecting any frictional torques: Regarding an instant t_0 just before the student pulls her limbs in, and an instant t_1 just after she finishes pulling her limbs in:

A. The angular momentum of the student plus that part of the stool that rotates is the same at t_0 as it is at time t_1.

B. The kinetic energy of the student plus that part of the stool that rotates is the same at t_0 as it is at time t_1.

C. Both A and B above.

D. None of the above.

6. A hollow sphere made of gold, and a solid sphere made of aluminum, both have the same mass and radius. These spheres are placed, side by side, at the top of an inclined plane and released from rest at the same time. How can we deduce that aluminum ball reaches the bottom of the incline first?

A. It has more rotational kinetic energy at the bottom of the incline.

B. It has less rotational kinetic energy at the bottom of the incline.

C. The premise of the question is flawed. The gold ball reaches the bottom of the incline first.

D. The premise of the question is flawed. Both balls reach the bottom of the incline at the same time.

7. A low-friction, freely-spinning, turntable with a mass M and radius R is spinning with an angular velocity ω, which points directly upward. A piece of putty, of mass m ($m < M$), is dropped so that it falls with a purely vertical velocity, v, and sticks to the outer edge of the turntable. The turntable slows down because the:

A. velocities of the putty and the turntable subtract due to vector addition.

B. moment of inertia of the rotating system increases.

C. angular momentum of the system decreases.

8. In a demonstration done in class, a professor, sitting on a stool that can rotate, put his feet on the ground, then holding a bicycle-wheel horizontal by its axel, started the wheel spinning. He then raised his feet and then flipped the wheel over so it was again horizontal but spinning in the opposite direction, as demonstrated in Figure 18.13.

FIGURE 18.13 Multiple choice question 8.

When this was done the professor started to spin in the direction the wheel was spinning before it was flipped. In this demonstration, the magnitude of angular momentum of the professor after the flip is greater than, less than, or equal to the magnitude of the angular momentum of the wheel before the flip?

A. greater than

B. less than

C. equal to

9. If a solid disk of mass (m) rolls without slipping with a constant linear velocity of (v) the total kinetic energy is:

A. $\frac{1}{2}m\,v_1^2$

B. $\frac{7}{10}m\,v_1^2$

C. $\frac{5}{6}mv_1^2$

D. mv_1^2

10. A uniform solid sphere has a mass M and a radius R, so it has a moment of inertia of I. If M is increased to $2M$ and R is increased to $3R$, what is the moment of inertia of the new sphere?

   A. I
   B. $2I$
   C. $6I$
   D. $12I$
   E. $18I$

18.8.2 PROBLEMS

1. The moment of inertia of the earth with respect to its axis of rotation is 8.04×10^{37} kg·m². Its rotation rate, relative to the distant stars, is once per 23.93 hours. Find the magnitude of the angular momentum of the earth, with respect to its axis of rotation, in kg·m²/s.

2. A skater is spinning at a rate of 2.00 rev/s on one skate with both arms and one leg extended. In this configuration, the skater has a moment of inertia of 3.00 kg·m². The skater pulls his legs and arms in, changing his moment of inertia to 1.00 kg·m². Find the spin rate (the magnitude of the angular velocity) of the skater after he has pulled his arms and legs in. Neglect air resistance and friction of the ice on the skater.

3. As described in Figure 18.6, a particle with a mass $m = 50$ g is moving with a velocity $v = 5$ m/s directly at point A, which is 25 cm from the center of the disk. The disk has a mass $M = 200$ g and a radius $R = 30$ cm and is mounted on a frictionless axis. If the particle, collides with the disk at point A and sticks to the disk, at what rate will the disk rotate after the collision?

4. At most bowling lanes, the bowling balls are sent back to the bowlers, and at the end, they roll up a ramp before they are stopped. If a bowling ball with a radius of 15 cm and a mass of 10 kg is rolled such that it rolls without slipping in the return gutter with a translational speed of 2.1 m/s, what translational speed will the ball have after it moves up the 30 cm high ramp? Consider the bowling ball to be a uniform solid sphere.

5. A spherically symmetric ball with a radius of 26 mm and a mass of 0.175 kg rolls without slipping down a straight incline. After the ball has rolled from rest at a height of 130 mm above a reference level, to the point where it is at the same elevation as the reference level, the translational speed of the center of the ball is 1.3 m/s. What is the moment of inertia of the ball?

6. A spring with a force constant of 6 N/m is mounted in a horizontal orientation, as described in Figure 18.14.

FIGURE 18.14 Problem 6.

One end of the spring is attached to a wall. The other end is not attached to anything. A marble, with a mass of 20 g and a radius of 1 cm, on a flat horizontal surface, is pushed up against the spring, compressing the spring 3 cm. The marble is released from rest and begins to roll without slipping on the surface. (Where it is in contact with the marble, the spring is frictionless.) Find the speed of the marble just after it leaves the spring, assuming that it continues to roll without slipping. Consider the marble to be a uniform sphere.

7. A disk with a mass of 5.0 kg and a radius of 50 cm is spinning at a constant rate of π rad/s and a mass of 1 kg is dropped onto the spinning disk, so that its center of mass is

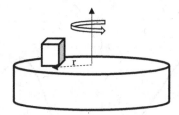

FIGURE 18.15 Problem 7.

at a distance $r = 30$ cm from the center of rotation of the disk, as described in Figure 18.15.

At what rate will the disk be spinning after the mass is dropped on the disk?

8. A neutron star is formed when a star, such as our sun, collapses under the force of its own gravity to one thousandth its original size. Suppose a uniform-spherical star of mass M and radius R collapses to a uniform spherical neutron star with a radius $\left(\frac{R}{1,000}\right)$, with the same mass. If the original star has a rotation rate of 1 rev in 25 days (like our sun), what will be the rotation rate of the neutron star in rev/days?

9. A uniform disk with radius (R) of 50 cm and a mass of 10 kg is located at the top of a ramp at rest, as depicted in Figure 18.16.

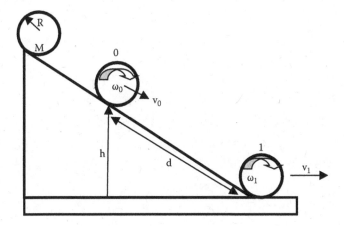

FIGURE 18.16 Problem 9.

The disk is released from rest from the top of the ramp, and the disk rolls without slipping. At a height of one half of a meter ($h = 0.5$ m) above a flat horizontal surface the disk, point 0, the disk is moving down the incline and spinning with an angular velocity of $\omega_0 = 2.2$ rad/s. The uniform disk continues to roll, without slipping, down the ramp and onto the horizontal surface. Compute the linear velocity (v_1) of the cm of the disk at the bottom of the ramp (point 1).

10. A hollow metal sphere, with a radius of 1 cm, has a mass of mass 50 g spread evenly around a uniform thin shell. This hollow thin sphere held at rest and located at the top of a ramp, at point A in Figure 18.17, is a height of 65 cm ($h_a = 65$ cm) above the horizontal level table.

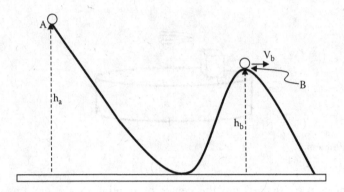

FIGURE 18.17   Problem 10.

The sphere rolls, without slipping, down the ramp and up and over a hill in the track. Point B is at height $h_b = 40$ cm above the level table top. Compute the linear velocity (v_b) of the sphere at point B as it rolls over the top of the hill.

19 Temperature and Heat

19.1 INTRODUCTION

With the conserved quantities of energy, momentum, and angular momentum established along with the process of quantity-based analysis, these concepts and techniques are first applied to thermodynamics. This chapter is dedicated to establishing the definition of terms employed in thermodynamics. Some of these terms may be familiar, and some may seem new, but all are important in thermodynamics. The chapter starts with a discussion about temperature and the three most common temperature scales. The terms internal energy and heat, along with the mechanism of heat transfer, are then discussed. Next, the caloric theory is reviewed, along with the process of analyzing the heat needed to change matter from one state to another. By the end of the chapter, you should have a good understanding of temperature, internal energy, heat, and some of the techniques used to compute these quantities.

19.2 CHAPTER QUESTION

Given the wide range of temperatures in the universe, why is the earth's average temperature about 57°F and the range in temperature on earth only about 262°F (−126°F to 136°F)? This question will be answered at the end of the chapter, employing most of the topics covered in the chapter.

19.3 TEMPERATURE, INTERNAL ENERGY, AND HEAT

As with work and energy, even though these words may sound familiar, they have specific definitions in physics. In thermodynamics, the words *temperature* and *heat* are familiar, but it is important to set a formal definition of each before applying them in analyzing a thermodynamic system.

19.3.1 Temperature (T°)

Temperature is a state variable, because it is used to define the state or condition of a system at a specific time. The other commonly employed state variables in thermodynamics are pressure (P) and volume (V), which were defined in previous chapters in volume 1. Temperature, which has obvious importance in thermodynamics, has the SI unit of the kelvin (K). Temperature is also an intensive variable, because a single value of temperature characterizes the entire system. That is, if half of the system is removed, the intensive variable still has the same value as the original system. In contrast, the value of an extensive variable changes with the quantity of the substance, so if half of the system is removed, the value of the variable changes. Mass is an example of an extensive variable. Consider the case of a gallon of milk in the refrigerator all at one temperature, assuming the container is not part of the system. If you pour half the milk out, the milk that remains has the same temperature, but it has half the mass of the original system.

Temperature is measured most easily by observing a physical characteristic of a system that depends on temperature. A device used to measure temperature is referred to as a thermometer. A common type of thermometer relates the length of a column of fluid confined in a tube to the temperature. Today, the most common liquid in such thermometers is dyed alcohol. The length of the column of fluid in a tube is calibrated to temperature by bringing the thermometer into contact with each of at least two systems, such as water at the freezing point and water at the boiling point, in both cases, at atmospheric pressure, at known temperatures.

DOI: 10.1201/9781003308072-20

The most common temperature scale calibration used in the United States is the Fahrenheit scale. Developed in 1724 by Daniel Fahrenheit, who set his 0° point to the temperature at which a salt water (brine) solution froze and his 100° point to the average human temperature. As you may know, the average human temperature today is 98,6°F.

The other common temperature scale is the Celsius scale, which was developed by Anders Celsius in 1742. For this scale, the 0°C and the 100°C are set at the melting point of ice and the boiling point of pure water (H_2O) at Standard Temperature and Pressure (STP), respectively. Because the set points are more easily repeatable, the Celsius scale is used more commonly in the rest of the world and in most scientific endeavors. The freezing and boiling points of water are used to develop conversion equations between the temperature scales. Given that water at STP freezes at 0°C and 32°F and boils at 100°C and 212°F, there are 100°C and 180°F between boiling and freezing of water. So, a Celsius degree, °C, is larger than a °F by the ratio of (180/100) = (9/5). Also, since 0°C and 32°F are the same physical point, there is an off-set by this amount. The conversion equations are as follows.

The conversion from Celsius measurements to Fahrenheit measurements is given in equation (19.1) as

$$°F = \left(\frac{9}{5}\right)°C + 32° \tag{19.1}$$

The conversion from Fahrenheit measurements to Celsius measurements is given in equation (19.2) as

$$°C = \left(\frac{5}{9}\right)(°F - 32°). \tag{19.2}$$

The Kelvin Temperature (K) scale is not commonly used in our daily life, but is important to the study of thermodynamics. This scale is important because the zero of this scale is at the point where all atomic and molecular motion stops, so it is the coldest theoretical temperate possible and is appropriately called absolute zero. This value of absolute zero is 0 K = −273.15°C. So, to convert from °C to K, just add 273.15, as shown in equation (19.3) as

$$K = °C + 273.15. \tag{19.3}$$

It is common not to use the degree symbol (°) with temperatures measured in the Kelvin scale. It should be noted that absolute zero has not yet been achieved experimentally, but scientists at different labs around the world have come close.

Since it is impossible to measure absolute zero with a conventional thermometer, it is measured through the relationships of the pressure (P) and volume (V) at known temperatures, and the ratios are extrapolated to absolute zero. For an ideal gas, the relationship between the state variables is ideal gas law, given in equation (19.4) as

$$PV = n R T, \tag{19.4}$$

where P is the pressure of the gas, V is the volume of the gas, n is the number of moles of the gas, R is the ideal gas constant $8.134\frac{J}{mole \cdot K}$, and T is the temperature of the gas in kelvin. The number of moles of a substance is given by a simple formula presented in equation (19.5) as

$$n = \frac{m(g)}{M\#}, \tag{19.5}$$

where ($M\#$) is the atomic mass of the atom, found in the periodic table, and $m(g)$ is the mass in grams. The number of molecules (or atoms) per mole is Avogadro's number, which is $6.02214 \times 10^{23}\frac{molecules}{mole}$.

Equations (19.5) is employed in the following examples:

Example 1. 1 g of hydrogen ($_1H^1$) results in $n = \frac{1g}{1} = 1$ *mole* of hydrogen and contains 6.022×10^{23} hydrogen atoms.

Example 2. 1 g of Carbon-12 ($_6C^{12}$) results in $n = \frac{1g}{12} = \frac{1}{12} = 0.0833$ *mole* of carbon and contains $(0.0833)(6.022 \times 10^{23})$ carbon atoms $= 5.01 \times 10^{22}$ carbon atoms.

Example 3. 1 g of Uranium-238 ($_{92}U^{238}$) is $n = \frac{1g}{238} = \frac{1}{238} = 0.0042$ *mole* of uranium and contains $(0.0042)(6.022 \times 10^{23})$ uranium atoms $= 2.53 \times 10^{21}$ uranium atoms.

19.3.2 Internal Energy (U)

Another important state variable is the **internal energy** (U) of a substance, which is the molecular-level energy, including both the random kinetic energy associated with the motion of the molecules, often referred to as thermal energy, and the potential energy due to the bonds between the molecules and/or atoms of a substance. The random kinetic energy does not include the kinetic energy associated with the motion of the entire object. Thus, the kinetic energy associated with the translational or rotational motion of an object as a whole, referred to as external energy when we want to contrast it with internal energy, is not included in the internal energy of the substance.

From a branch of physics known as statistical mechanics, there exists for an ideal gas a relationship between the average internal energy of the gas per atom and the temperature of the gas in kelvin. This relationship is given in equation (19.6) as

$$U_{ave} = \frac{3}{2}k_B T \tag{19.6}$$

where k_B is the Boltzmann constant, which has a value of 1.38065×10^{-23} J/K. The average internal energy is the energy per atom of the material. Note that this implies that at $T = 0$, the average translational kinetic energy of the atoms or molecules making up the system is zero, and that means that the average speed of a particle is zero. This is true in the classical physics model, which is in good agreement for most of the analysis done at normal temperatures, but in the very-low temperature range near absolute zero, quantum mechanics prevails, and the matter begins to behave in quantized ways. Fortunately, this is in a temperature region that is not common to any conditions that exist in nature on earth, so it is not something that is of concern for most common thermodynamic analysis.

19.3.3 Heat (Q)

Heat (Q) is another familiar term that is central to the study of thermodynamics. In thermodynamics, heat is defined as the transfer of thermal energy from one substance to another. Heat naturally moves from objects at a higher temperature to those at a lower temperature. The natural flow of heat raises the temperature of the cooler object and lowers the temperature of the hotter objects, until the two objects come to the same temperature, at which point the objects are said to be in thermal equilibrium.

19.3.4 Mechanisms of Heat Transfer

The three mechanisms for heat transfer from a hot substance to a colder one are conduction, convection, and radiation.

Conduction is heat transfer by molecular collisions. If one end of a metal rod is at a higher temperature (T_H) and the other end is at a lower temperature (T_L), as shown in Figure 19.1, then energy will be transferred from the hotter end to the colder end by conduction.

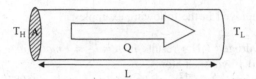

FIGURE 19.1   Flow of heat by conduction.

The particles moving at a higher speed at the hotter end will collide with slower ones, thus transferring energy and momentum to the slower molecules. This will continue to progress down the length, L, of the rod until, on average, all the particles are vibrating at the same speed, indicating a uniform temperature for the rod. This mechanism of heat transfer is common for solids and can be expressed in equation (19.7) as

$$Q = \frac{\kappa A (\Delta T) t}{L},\tag{19.7}$$

where Q is the heat, κ is the thermal conduction of the material, A is the area of the object, $\Delta T = T_H - T_C$ is the temperature difference from one end to the other end of the object, t is the elapsed time, and L is the length of the object. The thermal conductivities of some common materials are given in Table 19.1.

TABLE 19.1
Thermal Conductivities of Common Materials

Material	Thermal Conductivity [J/(m s K)]
Air at 0°C	0.024
Copper	385.0
Aluminum	205.0
Glass	0.8
Gold	310
Styrofoam	0.01

Conduction Example

Compute the heat flow down an aluminum rod that is 50 cm long, has a cross-sectional area of 2 cm^2 ($2 \times 10^{-4} \text{ m}^2$), and is kept at 100°C at the hot end and 20°C on the cold end.

Solution

Since the temperatures are given in °C, the change in temperature in °C is the same as that in K, so $\Delta T = 80$ K.

Starting with equation (19.7), plugging in the information gives,

$$Q = \frac{\kappa A (\Delta T) t}{L} = \frac{\left(205.0 \frac{J}{m \, s \, K}\right)(2 \times 10^{-4} m^2)(80 K)}{0.5 m} = 6.56 \, J$$

Convection is heat transfer by mass motion of a fluid, such as air or water, when the heated fluid is caused to move away from the source of heat, carrying energy with it. Convection above a hot surface occurs because hot air expands, becomes less dense, and rises. The cold air sinks, and this movement of warm air from the surface is convection. As you may gather, convection is one of the fundamental mechanisms in the working of the earth's atmosphere on a local, regional, and global scale. Convection is responsible for the formation of sea breezes on a hot summer day, and in combination with the earth's rotation, it is responsible for the global wind patterns. Convection is such a complicated fluid dynamic system that it does not lend itself to a simple equation and in fact is often a chaotic system.

Radiation is heat transfer by electromagnetic waves, such as light. This is how heat is transferred from the sun to the earth and from stars throughout the universe. The heat flow due to radiation can be expressed in formula form by the Stefan-Boltzmann Law, expressed in equation (19.8) as

$$Q = \sigma \varepsilon A T^4 t, \tag{19.8}$$

where σ is Stefan's constant (5.67×10^{-8} W/(m^2 K^4)), A is the area, T is the temperature in kelvin, t is the elapsed time, and ε is the emissivity, which measure how well an object absorbs or reflects radiation. An object that is a perfect reflector has a $\varepsilon = 0$, and an object that is a perfect absorber or emitter, which is called a blackbody, has an e = 1. Most real objects have an emissivity between the two extremes.

Radiation Example

Compute the heat flow per second by radiation from a light bulb filament at a temperature of 2928 K, an emissivity of 0.4, with an area of 60 mm^2 from the filament.

Solution
Convert the area of the filament from 60 mm^2 = 6 $\times$ 10^{-5} m^2.

Starting with equation (19.8), plugging in the information from the problem gives,

$$Q = \sigma \varepsilon A T^4 t = \left(5.67 \times 10^{-8} \frac{J}{m^2 \, s \, K^4} \right)(0.4)(6 \times 10^{-5} m^2)(2928 \, K)^4(1 \, s) = 100.0 \, J$$

Since a Joule per second is a Watt, this is a 100 W light bulb.

A simple **conceptual example** can help make it clear how each of the terms described so far in the chapter is used to describe the thermodynamic process. In a room, there is a 5 gallon pail full of room-temperature (20°C) water and a red-hot (400°C) penny. It is clear the red-hot coin is at a higher temperature than the water, so the penny has a much higher average internal energy than the water. But the water has more internal energy than the penny. Although the average molecules in the water are moving slower than those in the penny, there are so many more molecules of water. Therefore, the total energy is larger. Even though the water has a higher internal energy, heat will flow from the penny to the water if the penny is dropped into the water.

The heat will flow from the penny to the water by radiation and conduction. Since the penny is red hot, it is radiating visible light, and if you held your hand over the penny, you would sense the infrared radiation with your skin. In addition, the faster vibrating molecules of the penny will collide with the slower water molecules, transferring energy by individual collisions, which is the process of conduction. As the water molecules speed up, they push against adjacent molecules and spread out; thus, the water next to the hot penny becomes less dense. The water closest to the penny will rise and more dense, cooler water will move into its place. This is the flow of energy away from the penny by convection. These processes will continue until the penny and the water reach the same temperature, which is the state of thermal equilibrium. This provides an opportunity to state the **Zeroth Law of Thermodynamics;** that is, objects in thermal contact will reach thermal equilibrium, which is the same temperature.

19.4   CALORIC THEORY OF HEAT

The most common ways of sensing the heat transfer between objects is through measurements of the changes in the temperature and/or changes in the state of the matter from solid to liquid to vapor, and vice versa. The measurements of these effects are recorded in the specific heat of the material and its heat of fusion and heat of vaporization.

The specific heat (**c**) is the amount of heat per mass required to raise the temperature of a material by 1°C. The relationship that includes the specific heat and expresses the connection between heat and temperature change is given in equation (19.9) as

$$Q = mc\Delta T. \tag{19.9}$$

This expression assumes the substance remains in the same state of matter during the temperature change.

The specific heat of water is 1 calorie/gram °C = 4,186 J/kg°C, which is higher than most common substances. As a result, water plays a very important role in temperature regulation of our planet. The specific heat capacity c is a property of the material like density. As such, the values of specific heat for various substances are listed in Tables 19.1 and 19.2.

TABLE 19.2
Specific Heats of Some Solids and Liquids at a Constant Pressure

Substance	$c_p \left[\frac{J}{kg\,°C} \right]$
Copper	330
Iron	450
Aluminum	699
Glycerin	2320
Water (Liquid)	4186
Ice (Solid Water)	2000

The specific heat capacity of a substance varies with temperature and pressure. The values given in Tables 19.1 and 19.2 correspond to specific heats at atmospheric pressure.

Specific Heat Example

A 200 g block of iron at 100°C is dropped into an insulated container filled with 500 g of liquid water at 20°C. What is the final temperate of the water and iron block?

Solution

The heat that is "flowing" out of the iron is the same heat that is "flowing" into the water.
Thus, the sum of the heats must add to zero:

$$Q_{H_2O} + Q_{Fe} = 0$$

Rearranging the equation:

$$Q_{H_2O} = -Q_{Fe}$$

Plugging in the expression of Heat, equation (19.9):

$$(m_{H_2O})(c_{H_2O})(T_f - T_{iH_2O}) = -(m_{Fe})(c_{Fe})(T_f - T_{iFe})$$

Solve for T_f: *(notice that the negative sign switches the temperatures on the right-hand side of the equation.)*

Distribute:

$$(m_{H_2O})(c_{H_2O})(T_f) - (m_{H_2O})(c_{H_2O})(T_{iH_2O}) = (m_{Fe})(c_{Fe})(T_{iFe}) - (m_{Fe})(c_{Fe})(T_f)$$

Group terms of T_f on the left side of the equation:

$$(m_{H_2O})(c_{H_2O})(T_f) + (m_{Fe})(c_{Fe})(T_f)) = (m_{H_2O})(c_{H_2O})(T_{iH_2O}) + (m_{Fe})(c_{Fe})(T_{iFe})$$

Factor out T_f from the left side of the equation:

$$\left[(m_{H_2O})(c_{H_2O}) + (m_{Fe})(c_{Fe})\right]T_f = \left[(m_{H_2O})(c_{H_2O})(T_{iH_2O}) + (m_{Fe})(c_{Fe})(T_{iFe})\right]$$

Solve for T_f:

$$T_f = =\left[(m_{H_2O})(c_{H_2O})(T_{iH_2O}) + (m_{Fe})(c_{Fe})(T_{iFe})\right]/\left[(m_{H_2O})(c_{H_2O}) + (m_{Fe})(c_{Fe})\right]$$

The specific heats of water and iron are 4,186 J/kg°C and 450 J/kg°C, respectively.
Plugging in the numbers:

$$T_f = [(0.5 \text{ kg})(4,186 \text{ J/kg°C})(20°C) + (0.2 \text{ kg})(450 \text{ J/kg°C})(100°C)]$$
$$/[(0.5 \text{ kg})(4,186 \text{ J/kg°C}) + (0.2 \text{ kg})(450 \text{ J/kg°C})]$$
$$T_f = 23.3°C$$

19.5 CHANGES IN STATE

There are situations in which a hot object is brought into contact with a colder object in which there is no temperature change even though heat flows from the hot object into the cooler sample, but the temperature of the colder object does not increase. This occurs when the colder sample is undergoing a change in state from a solid to a liquid or a liquid to a vapor.

For instance, if a container filled with ice and liquid water at 0°C is brought in contact with a hot object, heat flows from the hot object into the ice water and the ice melts with no change in temperature. This will continue until all the ice is melted, assuming enough heat flows into the sample to melt all the ice. Then, after the last bit of ice melts at 0°C, if heat continues to flow into the water, the temperature of the water will increase.

This assumes the ice water is mixing away from the hot object so that the hot object is constantly in contact with some ice. The process described here is melting of ice, but it is similar to the freezing process. If water is put into an ice tray and then into the freezer, the cold air of the freezer comes in contact with the water and extracts heat from the water in the ice tray. The water in the ice tray cools from its initial temperature, probably around 55°F (12.8°C) to 32°F

(0°C). Then, as more heat flows from the water at 32°F (0°C), the water changes state from a liquid to a solid.

The other change of phase that is part of our common experience is vaporization and condensation. Again, if tap water initially at 55°F (12.8°C) is put into a pan at the same temperature and then put on the stove, the water will be heated. When the water reaches a rolling boil, the water is well mixed at 212°F (100°C), and the water molecules leave the water as water vapor. This is the process of vaporization. In this case, the liquid water remaining in the pot will remain at 212°F (100°C) as it boils. In fact, the vaporization process is maintaining the water at a constant temperature even as heat is added. The opposite process of vaporization is condensation, in which a gas is converted into a liquid.

In the situations described in the previous paragraphs, only the average state of the entire sample of water is the focus and not the transient, small in both time and space, changes in the water. It is true that water closest to the hot object or cold air in the freezer will get a bit hotter or colder and move away from the source of heat or sink, then return to the average temperature of the system. It is possible to analyze these transient changes in temperature across a substance with a differential equation called the heat equation, but it is beyond the mathematical scope of this text. So, in our analysis, we will concentrate on the overall changes in the system.

The quantity associated with the heat-per-mass needed to change the state of a substance is the *latent heat* (L). When the substance is freezing or melting, this quantity is known as the latent heat of fusion (L_f), and when the liquid is vaporized or a gas is condensed, the quantity is called the latent heat of vaporization (L_v). In terms of the latent heat, the amount of heat Q that must flow into a sample of a single-substance solid that is at the melting or vaporization (boiling) temperature is given by equation (19.10) as

$$Q = mL.$$ (19.10)

where the L is either L_f or L_v depending on the situation. There is no ΔT in the expression because there is no temperature change in the process. The whole phase change takes place at one temperature.

Tables 19.3 and 19.4 provide a list a few materials with the temperatures at which the transitions occur and the latent heat needed to change the state of the substance (Table 19.5).

TABLE 19.3

Specific Heats of Some Common Gases at Constant Pressure (c_p) and at Constant Volume (c_V)

Substance	$c_p \left[\frac{J}{kg \cdot °C} \right]$	$c_V \left[\frac{J}{kg \cdot °C} \right]$
Air	1003	717
Carbon Dioxide	842	653
Methane	2254	1735
Oxygen	922	662
Steam (Water Gas)	2000	1500

TABLE 19.4

Melting Points and Latent Heat of Fusion

Material	Melting Temperature (°C)	Latent Heat of Fusion (kJ/kg)
Helium	−269.65	5.23
Nitrogen	−209.97	25.5
Oxygen	−218.79	13.8
Water	0	334
Gold	1063	64.5
Copper	1083	134

TABLE 19.5

Boiling Points and Latent Heat of Vaporization

Material	Vaporization Temperature (°C)	Latent Heat of Vaporization (kJ/kg)
Helium	−268.93	20.9
Nitrogen	−195.81	201
Oxygen	−182.97	213
Water	100	2256
Gold	2660	1578
Copper	2567	5069

Heat of Fusion Example 1

Compute the heat needed to turn a 1.0 kg block of ice at 0°C to water vapor at 100°C.

Solution

To melt the 1.0 kg block of ice into liquid water it would take:

$$Q_m = m \ L_f = (1 \ \text{kg})(334 \ \text{kJ/kg}) = 334 \ \text{kJ}.$$

To raise the temperature of the water from 0°C to 100°C, it would take:

$$Q_T = m \ c \ \Delta T = (1 \ \text{kg})(4,186 \ \text{J/kg°C})(100°C) = 418.6 \ \text{J} = 0.4186 \ \text{kJ}$$

To vaporize the 1.0 kg of water, it would take:

$$Q_V = m \ L_v = (1 \ \text{kg})(2256 \ \text{kJ/kg}) = 2256 \ \text{kJ}.$$

So, the total amount of heat needed to melt a 1.0k g piece at 0°C of ice into 1.0 kg of steam at 100°C is:

$$Q = Q_m + Q_T + Q_V = 334 \ \text{kJ} + 0.4186 \ \text{kJ} + 2256 \ \text{kJ} = 3008.6 \ \text{kJ}$$

Heat of Fusion and/or Heat of Melting Example 2

A 2.0 g piece of ice is dropped into an insulated cup containing 100 g of liquid water at 20°C. What is the final temperature of the resulting water?

Solution

The heat from the water must go into melting the ice and heating the ice water from 0° to the final temp:

$$[Q_I + Q_{IW}] + Q_w \qquad\qquad = 0.$$
$$[m_I L_f + m_I c\Delta T] + m_w c\Delta T = 0$$

$[m_I L_f + m_I c(T_f\ T_{Ii})] = -m_w c\ (T_f - T_{wi})$ Heat into the ice and ice water and from the water
$[(m_I c + m_w c)\ T_f] = m_w c T_{wi} + m_I c\ T_{Ii} - m_I\ L_f$ (T_f is the same for all the water)

$$T_f = [m_w c T_{wi} + m_I c T_{Ii} - m_I L_f]/[(m_I + m_w)c]$$

$$T_f = [(100g)(4.186\ \text{J/g°C})(20°C) + (2g)(4.186\ \text{J/g°C})(0°C) - (2g)(334\ \text{J/g})]$$

$$/[((2 + 100)g)(4.186\ \text{J/g°C})]$$

$$T_f = 18.0°C$$

19.6   ANSWER TO THE CHAPTER QUESTION

With the information from this chapter, an answer to the chapter question about the temperature of the earth can be developed. The main source of energy for the earth's environment is the sun, and radiation is the mechanism by which energy from the sun is delivered to the earth. Therefore, using the Stefan-Boltzmann Law, equation (19.8), with the quantities specific to the Sun, the energy per second [$t = 1$ s] can be computed assuming the values of: σ is Stefan's constant [5.67×10^{-8} W/(m^2 K^4)], ε is the emissivity of the sun [0.99], A is the area of the sun [6.07×10^{18} m^2], T is the temperature of the surface of sun [5800 K],

$$Q = \sigma\varepsilon A T^4 t = \left(5.67 \times 10^{-8}\frac{W}{K^4 m^2}\right)(0.99)(6.07 \times 10^{18}m^2)(5800K)^4(1\ s)$$

$$Q = 3.86 \times 10^{26} Ws = 3.86 \times 10^{26} J, \text{ every second}$$

Since the earth is 149.6 billion meters from the sun and the earth has a radius of 6.371 million m and the area of ½ of a sphere is ½ ($4\pi r^2$), which is the shape of the earth that intercepts the solar radiation, the fraction of the overall solar radiation that is intercepted by the earth is

$$fraction\ int = \frac{Area\ of\ \frac{1}{2}the\ earth\ surface}{Area\ of\ radiated\ solar\ radiation\ at\ the\ distance\ from\ the\ sun\ to\ the\ earth}$$

$$fraction\ int = \frac{\frac{1}{2}(4\pi(6.371 \times 10^6 m)^2)}{4\pi(1.496 \times 10^{11}m)^2} = 1.81 \times 10^{-9}$$

So, the total heat delivered to the earth due to solar radiation is the product of fraction of radiation intercepted and the total radiated by the sun.

$$Q = (3.86 \times 10^{26} Ws)(1.81 \times 10^{-9}) = 6.99 \times \frac{10^{17}J}{s} = 6.99 \times 10^{17}W.$$

About 30% of this radiation is reflected before it is absorbed by the earth's surface and atmosphere, so the total heat from the sun absorbed by the earth is

$$Q = (0.7)6.99 \times 10^{17}W = 4.74 \times 10^{17}W.$$

This is the heat per second upon which the entire earth system operates. It must be true that the amount of heat that escapes the earth must be equal to the amount that enters into the earth system or it would continue to get hotter and hotter and not be in equilibrium.

The key to the earth maintaining a fairly constant temperature is the time delay of the processes by which the radiation from the sun makes its way through the earth system. This incoming solar radiation heats up the surface of the planet, and the surface heats up the atmosphere by several mechanisms, radiation and convection. It is the convection of air and water in the earth system that drives our weather. Water evaporated off the oceans and lakes and fresh water rises into the atmosphere. As it rises it cools, eventually condensing into clouds, which are actually suspended liquid water droplets. These processes transport heat from the surface to the atmosphere through many of the mechanisms discussed in this chapter.

19.7 QUESTIONS AND PROBLEMS

19.7.1 MULTIPLE CHOICE QUESTIONS

1. Is temperature an intensive or an extensive variable?
   A. intensive
   B. extensive
2. Which temperature scale is set with the 0° and the 100° at the melting point of ice and the boiling point of pure water (H_2O) at Standard Temperature and Pressure (STP).
   A. Fahrenheit
   B. Celsius
   C. Kelvin
3. Which system has a greater average internal energy, a system that is at 20 K or a system that is at 20°C?
   A. the system at 20 K
   B. the system at 20°C
   C. neither – both are the same temperature
4. Which system experiences the greater increase in its average internal energy, a system whose temperature increases by 20 K or a system whose temperature increases by 20°C?
   A. the system whose temperature increases by 20 K
   B. the system whose temperature increases by 20°C
   C. neither – both experience the same increase in average internal energy
5. Which system experiences the greater increase in its average internal energy, a system whose temperature increases by 20°F or a system whose temperature increases by 20°C?
   A. the system whose temperature increases by 20°F
   B. the system whose temperature increases by 20°C
   C. neither – both experience the same increase in average internal energy
6. The internal energy (U) of the working substance includes the translational and rotational motion of an object as a whole.
   A. true
   B. false
7. 4 kg of water at 100°C is poured into an insulated cooler containing 1 kg of water at 0°C. Neglecting the cooler, what will be the equilibrium temperature of the total amount of water?
   A. 0°C
   B. 20°C
   C. 50°C
   D. 80°C
   E. 100°C

8. A 1.0 kg block of copper (c_{Cu} = 386 J/kg°C) at a temperature of 0°C is put into an insulated cooler containing 1 kg of water (c_{H2O} = 4186 J/kg°C) at 100°C. Neglecting the cooler, what is the final equilibrium temperature of the water and copper block?
   A. 0°C
   B. between 0°C and 50°C
   C. 50°C
   D. between 50°C and 100°C
   E. 100°C

9. 1.0 g of ice at 0°C is placed with 10.0 g of water at 50°C. What is the final temperature of the total amount of water? (L_F = 80 cal/g and c_{water} = 1 cal/g°C)
   A. 0°C
   B. between 0°C and 50°C
   C. 50°C

10. 1 kg of water at 100°C is poured into an insulated cooler containing 1.0 kg of water at 0°C. Neglecting the cooler, what will be the equilibrium temperature of the total amount of water?
   A. 0°C
   B. 20°C
   C. 50°C
   D. 80°C
   E. 100°C

19.7.2 PROBLEMS

1. Considering 70°F to be room temperature, calculate room temperature in °C and in kelvin.
2. According to a 2020 *Scientific American* article, the average internal temperature of the American human body is now 97.5°F. Calculate the current average internal temperature of the American human body in °C and in kelvin.
3. Calculate absolute zero (0 K) in degrees Fahrenheit (°F).
4. Compute the number of atoms in 1.0 g of Uranium-238 ($_{92}U^{238}$).
5. Calculate the average internal energy of an ideal gas at 20°C.
6. Compute the heat flow per second down a copper rod that is 20 cm long, has a cross-sectional area of 3 cm^2, and the hot end is at 80°C and the cold end is kept at 30°C.
7. Compute the heat flow per second by radiation from a star with a surface temperature of 5000 K, an emissivity of 0.99, with a surface area of 9 × 10^{18} m^2.
8. Compute the heat needed to raise the temperature of a 200 g piece of iron from 30°C to 50°C.
9. Calculate the specific heat of a metal from the following data: A container made of the metal has a mass of 3.6 kg and contains 14.0 kg of water. A 1.8 kg piece of the metal initially at a temperature of 180°C is lowered into the water and released. The container and water initially have a temperature of 16°C, and the final temperature of the entire system is 18°C.
10. What mass of steam at 100°C must be mixed with 150 g of ice at 0°C, in a thermally insulated container, to produce liquid water at 50°C?

20 Thermodynamics of Heat Engines

20.1 INTRODUCTION

This chapter begins with the first law of thermodynamics, which is a statement of conservation of energy. The exchange of heat, internal energy, and work is first applied to a simple system comprised of a cylinder filled with an ideal gas and fitted with a piston. This system becomes the standard in which different thermodynamic processes, including those in which state variables such as pressure, temperature, and volume, are kept constant as the other quantities change. These processes are linked together in a way that the system returns to its initial condition so it can be called a cycle. Several examples of well-known cycles are presented, along with the PV-diagram that is central to the analysis of these systems. The Carnot cycle is introduced as the most efficient cycle possible, and this leads to the concept of entropy as a transition to the second law of thermodynamics.

20.2 CHAPTER QUESTION

Did you ever wonder why it seems hot to us when it is 98°F outside, even though that is about the same temperature that our bodies maintain? That seems a bit of a mystery, along with the question of why we seem to prefer temperatures of around 70°F, even though this is more than 25°F below our body temperature. The first law of thermodynamics and the study of heat engines will provide context to answer these questions at the end of the chapter.

20.3 FIRST LAW OF THERMODYNAMICS

The first law of thermodynamics is an expression of conservation of energy applied to the concepts of heat (Q), internal energy (U), and work (W). The discussion of this law will start with a simple piston-cylinder arrangement, depicted in Figure 20.1.

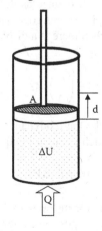

FIGURE 20.1 Basic piston-cylinder arrangement.

This system consists of a piston, with a cross-sectional area (A) that is tightly fit into a cylinder. The piston is able to slide up and down in the cylinder without friction. The cylinder is filled with a volume (V) of an ideal gas. If heat (Q) is added to a system, the heat will either change the internal energy (ΔU)

DOI: 10.1201/9781003308072-21

of the gas by speeding up or changing the grouping of the atoms of the gas and/or it will do work *(W)* on the system. An equation form of the first law of thermodynamics is given in equation (20.1) as,

$$Q = \Delta U + W \tag{20.1}$$

In the previous chapter, the quantities of internal energy (*U*) and heat (*Q*) were introduced, and examples were done for each. Remember that there are expressions to find the heat needed to raise the temperature of an object, $Q = mc\Delta T$, or to melt or vaporize an object, $Q = mL$. Calculations were done to find the amount of heat transferred through conductivity, $Q = \frac{\kappa A (\Delta T) t}{L}$, and radiation, $Q = \sigma \varepsilon A T^4 t$. In addition, there are expressions to find the internal energy, per atom, for an ideal gas at a given temperature, $U_{ave} = \frac{3}{2} k_B T$. In practice, while working with the first law of thermodynamics, equation (20.1), the heat added or subtracted from a system can be found in all the ways introduced in the previous chapter, but the total value of the internal energy is commonly not of interest, and only the change in internal energy (*ΔU*) is of interest. The other term in the first law of thermodynamics is the work done on or by the gas.

20.3.1 WORK (W)

The substances in most thermodynamic systems that do work, or have work done on them, are commonly gases, which will be approximated as an ideal gas. The work done on or by a thermodynamic system on these gases is commonly measured in terms of the pressure and volume. Given the definition of the work in equation (16.1) as

$$W = Fd\cos(\theta_F - \theta_d),$$

which is the product of the force (*F*) on the piston, the distance the piston rises (*d*), and the cosine of the difference in the angle between the force and displacement done on the gas. The work on the system, shown in Figure 20.1, reduces to

$$W = Fd,$$

because the angles of the force and displacement are the same for the piston-cylinder system and the $\cos(0°) = 1$. In addition, since pressure is defined as the force per area in equation (10.1) and the product of the displacement of the piston and cross-sectional area (*A*) is the change in the volume (*ΔV*) of the gas in the cylinder, when there is a displacement (d) of the work on the piston, the equation is:

$$W = \left(\frac{F}{A}\right)(Ad).$$

This leads to the expression for work in the thermodynamic system as the product of the pressure and the change in volume of the gas, which is given in equation (20.2) as,

$$W = P(\Delta V) \tag{20.2}$$

This expression not only gives the magnitude of the work, but it also gives the sign of the work. Remember, work is not a vector, but it can be positive or negative. In the thermodynamic system in Figure 20.1, positive work occurs when the gas expands, so the gas does work in lifting the piston, and negative work occurs when the gas is compressed, and the piston does work on the gas. In addition, if the pressure is given in units of Pascals (Pa) and the change in volume is given in cubic meters (m³), the product of *P(ΔV)* results in work in Joules,

$$(Pa)(m^3) = \left(\frac{N}{m^2}\right)(m^3) = Nm = J$$

20.4 HEAT ENGINE

A heat engine is a device that converts heat to work, and in the process, it may convert some of the heat to a change in internal energy. The heat input can be many different forms: from a flame, to a change of state from gas to liquid, or even a chemical reaction like the combustion of gasoline. The first law of thermodynamics is used to explain each step and the complete cycle of the heat engine. Each step in the cycle can be one of the four types of thermodynamic processes, which are:

1. **Isochoric** is a process in which the volume is kept constant. Therefore, the change in volume is zero, $\Delta V = 0$, so $W = P(\Delta V) = P(0) = 0$, so by the first law of thermodynamics: $Q = \Delta U$,
2. **Isobaric** is a process in which the pressure is kept constant. Therefore, the work is given by equation (20.2) as $W = P(\Delta V)$, so the first law is $Q = P\Delta V + \Delta U$,
3. **Isothermal** is a process in which the temperature is kept constant, so by equation (19.8) $U_{ave} = \frac{3}{2}k_B T$, there is no change in internal energy, so $\Delta U = 0$. Therefore, the first law gives $Q = W$. So, the work done on or by the gas is equal to the heat input of output of the gas. If the heat flow can be found, the work is the same value. To find the work done during an isothermal process, use equation (20.3):

$$W = nRT \ln\left(\frac{V_f}{V_i}\right), \tag{20.3}$$

where the temperature T must be in kelvin, V_f is the final volume, and V_i is the initial volume of the process, respectively. The derivation of equation (20.3) requires the use of integral calculus to generate the natural log, but hopefully it is apparent that the substitution for P in the expression of work in equation (20.2) accounts for the nRT that comes from the ideal gas law along with a $\left(\frac{\Delta V}{V}\right)$ that is key to producing the natural log.

4. **Adiabatic** is a process in which heat is kept constant; none is added or subtracted from the system, so $Q = 0$, so $\Delta U = -W$.

The analysis techniques for heat engines will be introduced with the simple four-step heat engine pictured in Figure 20.2.

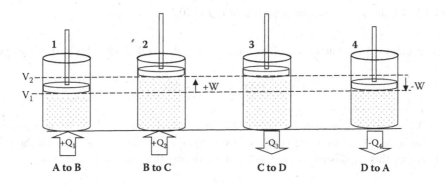

FIGURE 20.2 The four steps in a heat-engine cycle.

Assume the piston is free to move in the cylinder and the cylinder is filled with an ideal gas to a volume V_1.

In step **1,** heat is added to the piston-cylinder while it is kept at a constant volume (V_1). Assume there is a latch on the piston so it can be held at volume (V_1), while heat (Q_1) is input into the system with something like a Bunsen burner. In step **2**, the latch is released and the piston is allowed to expand at a constant pressure (P_2), as more heat (Q_2) is added. No restriction is needed to have the piston move at a constant pressure, since the gas will expand until it reaches equilibrium, continually achieving a constant pressure. In step **3**, heat (Q_3) is extracted from the system at a constant volume (V_2). Assume a second latch is used to keep the piston at a constant volume of V_2. In step **4**, the gas compresses at a constant pressure (P_1), while more heat (Q_4) is extracted, so the gas returns to the original volume and back to the original state.

20.4.1   PV-Diagrams

A common graphical device used to keep track of the operation of the cycle is a pressure volume diagram or just PV-diagram for short. A PV-diagram for the cycle presented in Figure 20.2 is shown below in Figure 20.3. Starting at point A, with a pressure (P_1) and a volume (V_1), heat (Q_1) is added in step 1, which takes the piston-cylinder to pressure P_2 at the same volume V_1. This process, in which the volume is constant, is called an **isochoric process**. In step 2, the piston is allowed to expand at a constant pressure of P_2 while heat (Q_2) is added. This process, in which the pressure is constant, is called an **isobaric process**. In step 3, heat (Q_3) is extracted while the volume is kept constant (V_2), so the pressure decreases to P_1. In step 4, heat (Q_4) is extracted, and the piston is allowed to contract to V_1, at a constant pressure P_1. The cycle is back at the starting point, ready to do it all over again.

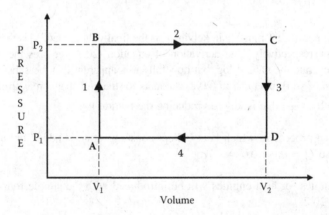

FIGURE 20.3   PV-diagram of the four-step cycle in Figure 20.2.

For this cycle, work is done in steps 2 and 4. In step 2, the work during this part of the cycle is,

$$W_2 = P\Delta V = P_2(V_2 - V_1).$$

Since $V_2 > V_1$, this work is positive, which indicates that the gas is doing work lifting the piston. If the piston has a mass (m), the work in this step should equal the work it takes to lift the piston a height (h) during this expansion. In step 4, the work during this part of the cycle is,

$$W_4 = P\Delta V = P_1(V_1 - V_2).$$

Since $V_2 > V_1$, this work is negative, which indicates that work is done on the gas compressing it due to the weight of the piston. Since there is no change in volume, and thus no displacement, in steps 1 and 3 no work is done. So, the net work done during this cycle is:

$$W_{net} = W_1 + W_2 + W_3 + W_4.$$

which is:

$$W_{net} = 0 + P_2(V_2 - V_1) + 0 + P_1(V_1 - V_2) = (P_2 - P_1)(V_2 - V_1).$$

Thus, the **work done** by the cycle is just the **area enclosed by the PV-diagram** of the cycle. This relationship is an important one, and it holds for all thermodynamic cycles. A cycle that uses heat to do work has a PV-diagram that moves clockwise around the diagram. A cycle that uses work to move heat in a direction it would not naturally go has a PV-diagram that moves counterclockwise around the diagram.

The two other types of processes, **isothermal** and **adiabatic**, are curves on the PV-diagram and depend on the type of gas law that best represents the working gas.

20.4.2 Gas Laws for Heat Engines

In most thermodynamic systems, such as engines, refrigeration systems, and heat pumps, the material that does the work or has the work done on it is a fluid. Most of the time, this fluid is a gas, and sometimes the fluid changes state between a gas and liquid. Even in these cases, most of the time the working fluid does work is when the fluid is in the gas state. Thus, gas laws play an important role in thermodynamics, providing the connections between the measurable state variables of the gas. As described in Chapter 19, the ideal gas law is a relation between P, V, and T, given in equation (19.4) as

$$PV = nRT.$$

Even if the gas is not an ideal gas, the temperature of the gas is usually related to the product of pressure and volume of the gas. For many real gases, the relationship between the pressure, volume, and temperature of the gas can be expressed as:

$$PV^\gamma = nRT,$$

which is known as a **polytropic gas law**. The value of γ changes for different types of gas, with values such as $\gamma = 1.4$ for air. So, for ideal and most real gases, the temperature of a gas is related to a point (P, V) on the PV-diagram. In addition, since the temperature of a gas is proportional to the internal energy of a gas, given in equation (19.8) as $U = \frac{3}{2}k_B T$ so each point of the PV-diagram is related to a specific value for the internal energy. That is, the pressure (P) and the volume (V) of a gas determine the temperature of the gas, which in turn determines the internal energy of the gas. Thus, since all cycles return back to the starting point, the net change in internal energy of the working substance, the gas, is zero. That is, for a complete cycle $\Delta U_{cycle} = 0$.

20.4.3 The Otto Cycle – An Example of a Heat Engine

One interesting example of a cycle with curved PV-lines is the Otto Cycle, pictured in Figure 20.4. This cycle is a good approximation of the gasoline combustion engine, like in many automobiles. The Otto Cycle is made up of isochoric and adiabatic transitions.

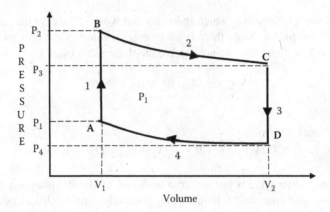

FIGURE 20.4   PV-diagram of the Otto Cycle.

In step **1,** heat is added to the piston-cylinder when the gasoline-air mixture explodes due to the spark from the sparkplug. This happens so quickly that the pressure increases while the volume (V_1) remains constant. In step **2**, the gas in the piston expands with no further change in the heat. In step **3**, the exhaust valve(s) open, and the pressure rapidly decreases as the hot exhaust escapes out of the cylinder. Again, like step 1, this step happens so fast that the volume of the cylinder lags behind the change in the pressure. In step **4**, the new gas-air mixture is pumped into the cylinder and is compressed with no heat input into the system. At point A, the sparkplug fires again, and the cycle repeats. It is amazing to think that this is happening 2000–5000 times per minute for each cylinder in a car. There are many different types of cycles for many different applications, but there is only one that is theoretically the most efficient.

20.4.4 EFFICIENCY

Any heat engine takes energy, in the form of heat, in from a hot thermal reservoir, converts some of that energy into work, and dumps some of that energy to a cold thermal reservoir (see Figure 20.5).

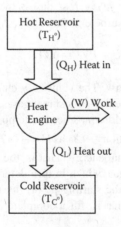

FIGURE 20.5   Energy flow diagram of a heat engine.

For a coal-fired steam engine, the hot thermal reservoir is the burning coal, and the cold thermal reservoir is the atmosphere. For a steam turbine that is part of a nuclear power plant at the coast, the hot thermal reservoir is a nuclear reactor, and the cold thermal reservoir is the ocean. The efficiency of a heat engine is the ratio of the work out to the energy that flows in from the hot reservoir in the form of heat. Remember that work is just the past of energy, so efficiency is a ratio of energies. The less energy the heat engine has to dump to the cold reservoir, the more efficient the engine. Let W_{OUT} be the net work out during a cycle. Let Q_H be the amount of heat that flows into the working fluid from the hot thermal reservoir, and let Q_C be the amount of heat that flows out of the working fluid to the cold thermal reservoir. The efficiency of any heat engine is given by equation (20.4) as,

$$\eta = \frac{W_{OUT}}{Q_H} \qquad (20.4)$$

The fact that the working substance returns to its original state at the end of each cycle, meaning there is no net change in the internal energy of the working substance, means that the work out is equal to the net amount of heat in, $W_{OUT} = Q_H - Q_C$. Thus, the efficiency for any heat engine can be expressed in terms of the heat exchange of the engine as equation (20.5), which is,

$$\eta = \frac{Q_H - Q_C}{Q_H} = 1 - \frac{Q_C}{Q_H} \qquad (20.5)$$

20.4.5   HEAT ENGINE EXAMPLES

Example 1

Like the piston-cylinder shown in Figure 20.1, an insulated ridged wall cylinder is fitted with a circular piston, which has an area of 2 m². A pressure sensor is used to monitor the pressure of the ideal gas in the cylinder. The atmospheric pressure on the day of this experiment is 100,000 Pa, and a block is then placed on the top of the piston so that the pressure of the gas in the cylinder due to the weight of the block is increased by 400 Pa and the piston is at rest sitting on a 1 m³ column of gas so the pressure sensor reads 100,400 Pa. The pressure and volume of the starting point of this process is labeled point A in the PV-diagram of Figure 20.6.

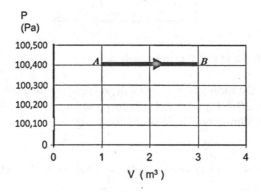

FIGURE 20.6   PV-diagram of a heat engine in example 1.

Then, 400,000 J of heat are added to the gas in the cylinder, and the piston rises so that the volume increases at a constant pressure, as described in the PV-diagram. What is the change in the internal energy of the gas during this expansion?

Solution

It is important to note that this process is isobaric, so the work done during the processes is given by equation (20.2) and the values from the PV-diagram:

$$W = P(\Delta V) = 100,400 \ Pa(3 \ m^3 - 1 \ m^3) = 100,400 \ Pa(2 \ m^3) = 200,800 \ J$$

From the first law of thermodynamics, given in equation (20.1) as $Q = \Delta U + W$, the change in internal energy can be found from the heat added, given in the problem as 400,000 J, and the work found in the previous step in the solution,

$$\Delta U = Q - W = 400,000 \ J - 200,800 \ J = 199,200 \ J$$

This is the answer to the problem. In addition, since the internal energy increases, the temperature of the ideal gas at point B is higher than it was at point A in the PV-diagram. Therefore, some of the heat went into increasing the average internal energy of the gas, and some went into doing work lifting the block and the piston. Since 400 Pa of pressure in the ideal gas is associated with the block and the area of the piston is 2 m², the weight of the block is $mgh = F = P(A) = 400 \ Pa(2 \ m^3) = 800 \ N$. Since the change in the volume is 2 m³ and the area of the piston is 2 m², the vertical displacement of the piston and block is $d = \frac{V}{A} = 1 \ m$. So, the change in the gravitational potential energy of the block is $mgh = (800 \ N)(1 \ m) = 800$. The other 200,000 J of work is done against the atmosphere lifting the piston $W = Fd = (PA)d = (100,000 \ Pa)(2 \ m^2)(1 \ m) = 200,000 \ J$.

Example 2

The PV-diagram for a well-insulated heat engine is shown in Figure 20.7.

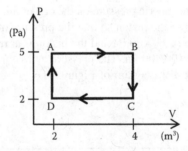

FIGURE 20.7   PV-diagram of a heat engine in example 2.

The internal energy of the ideal gas used in the heat engine at the points at the four corners of the PV-diagram are, $U_A = 20 \ J$, $U_B = 40 \ J$, $U_C = 16 \ J$ and $U_D = 8 \ J$. Find the work done and the heat added and subtracted during one cycle of this heat engine.

Solution

First, find the work done in each step of the cycle and total work done in the entire cycle.

$$
\begin{aligned}
W_{AB} &= P(\Delta V) = 5 \ Pa(4 \ m^3 - 2 \ m^3) = 10 \ J \\
W_{BC} &= P(\Delta V) = (5 \ to \ 2) \ Pa(0 \ m^3) = 0 \ J \\
W_{CD} &= P(\Delta V) = 2 \ Pa(2 \ m^3 - 4 \ m^3) \ = -4 \ J \\
W_{DA} &= P(\Delta V) = (2 \ to \ 5 \ Pa)(0 \ m^3) = 0 \ J \\
W_{total} &= W_{AB} + W_{AB} + W_{CD} + W_{DA} = 6 \ J
\end{aligned}
$$

Or, the area enclosed by the PV-diagram

$$W_{area} = \Delta PDV = (5 \ Pa - 2 \ Pa)(4 \ m^3 - 2 \ m^3)$$
$$W_{area} = (3 \ Pa)(2 \ m^3) = 6 \ J$$

Next, find the change in internal energy for each step.

$$\Delta U_{AB} = U_B - U_A = 40 \ J - 20 \ J = 20 \ J$$
$$\Delta U_{BC} = U_C - U_B = 16 \ J - 40 \ J = -24 \ J$$
$$\Delta U_{CD} = U_D - U_C = 8 \ J - 16 \ J = -8 \ J$$
$$\Delta U_{DA} = U_A - U_D = 20 \ J - 8 \ J = 12 \ J$$
$$U_{total} = U_{AB} + U_{AB} + U_{CD} + U_{DA} = 0 \ J.$$

This a good check since the change in internal energy should be zero for any closed cycle.

To find the heat input for each step use the first law of thermodynamics for each step in the cycle:

$$Q_{AB} = W_{AB} + \Delta U_{AB} = 10 \ J + 20 \ J = 30 \ J$$
$$Q_{BC} = W_{BC} + \Delta U_{BC} = 0 \ J + -24 \ J = -24 \ J$$
$$Q_{CD} = W_{CD} + \Delta U_{CD} = -4 \ J + -8 \ J = -12 \ J$$
$$Q_{DA} = W_{DA} + \Delta U_{DA} = 0 \ J + 12 \ J = 12 \ J$$
$$Q_{total} = Q_{AB} + Q_{BC} + Q_{CD} + Q_{DA} = 8 \ J$$

The efficiency of the cycle can be found in two ways, first by dividing the work done by one cycle of the ideal gas, which is 6 J, and the heat into the system, which is the sum of Q_{AB} and Q_{DA}, which is 30 J + 12 J = 42 J. The other way is to divide the difference in the heat in to the heat out by the heat into the system.

$$\eta = \frac{W_{OUT}}{Q_H} = \frac{6 \ J}{30J + 12J} = \frac{Q_H - Q_C}{Q_H} = \frac{42 - 36}{42} = 0.143 = 14.3\%$$

Both calculations give an efficiency of approximately 14.3%.

20.4.6 CARNOT CYCLE

Theoretically, the upper limit of heat-engine efficiency is not 100% but the efficiency of the Carnot cycle. Developed by Nicolas Carnot in 1824, the Carnot cycle is considered the most efficient cycle possible because the exchanges of heat are done isothermally, after which all available heat is used to continue to do work adiabatically. Consider a piston-cylinder arrangement, like in Figure 20.1, in which a cylinder capped by a piston is filled with an ideal gas and heat is added to and subtracted from the ideal gas using a thermal reservoir, which is an object or substance that heat flows into or out of the working substance, the ideal gas in this case, of the heat engine and there is no temperature change of the thermal reservoir. This is possible if the thermal reservoir is large compared to the heat engine, like the ocean to a nuclear reactor. Thus, by definition, a thermal reservoir is always at one constant temperature. In the Carnot cycle, heat is added isothermally, so all available energy is converted to work. Then, any remaining energy is converted to work adiabatically, the exhaust is handled isothermally to avoid any loss of heat from the ideal gas, and finally the system returns to the starting point adiabatically to avoid any loss of heat to the surroundings. Remember that when the piston moves upward, the gas does work on the surroundings, and when the gas contracts, the piston moves downward and the surroundings do work on the gas. The details of the Carnot cycle are most easily explained with the aid of a PV-diagram, shown in Figure 20.8.

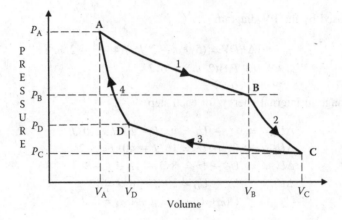

FIGURE 20.8   PV-diagram of the Carnot cycle.

The following section is a step-by-step explanation of the Carnot cycle.

In *step 1* (A → B), the cylinder of gas is in contact with the hotter thermal reservoir, and heat flows into the ideal gas, the working substance of the Carnot engine isothermally. An isothermal process is one that takes place with no change in temperature, at temperature T_1 so all the heat input into the gas goes into work done by the gas. The gas expands, doing work on the surroundings. The work done by the gas is computed with equation (20.3), $W_{AB} = nRT_1 \ln(V_B/V_A)$, with T_1 in kelvin. The fact that the temperature does not change during this step means that the internal energy of the ideal gas does not change. Given that the internal energy does not change, all the energy that enters the ideal gas in the form of heat in this step, leaves the ideal gas in the form of work in this step so $Q_{AB} = W_{AB}$.

In *step 2* (B → C), the ideal gas expands adiabatically, an adiabatic process is one in which there is no heat flow into or out of the system. Since $Q_{BC} = 0$, $W_{BC} = -\Delta U_{BC}$, the expanding gas does work on the surroundings, so the internal energy decreases. That is why ΔU_{BC} is negative, meaning the temperature of the gas decreases. The temperature decreases to T_3, the temperature at which step 3 takes place.

In *step 3* (C→D), the cylinder is in contact with the colder thermal reservoir, so heat flows out of the gas isothermally at a temperature T_3. The amount of heat flowing out in this step is Q_{CD}. The gas is being compressed, meaning that the surroundings are doing work on the gas. As in step 1, the pressure is varying, but throughout this step, the pressure is lower than the lowest pressure in step 1, so less work is done on the gas in this step than the gas did on the surroundings in step 1. This work, again calculated using equation (20.3), is $W_{CD,3} = nRT_3 \ln(V_C/V_D)$, with T_3 in kelvin. The fact that the temperature does not change during this step means that the internal energy of the ideal gas does not change. Given that the internal energy does not change, all the energy that enters the ideal gas in the form of work in this step, leaves the ideal gas in the form of heat in this step. $Q_{CD,3} = W_{CD,3}$.

In *step 4* (D → A), the surroundings compress the gas adiabatically. Since $Q_{DA,4} = 0$, $W_{DA} = \Delta U_{DA}$, and the fact that the gas is being compressed means that work is done on the gas and the internal energy increases, so the temperature of the gas increases. The temperature increases to T_1, the temperature at which step 1 takes place. Upon completion of step 4, the working substance is back in its original state, ready for another cycle.

In steps 1 and 2, work is done by the ideal gas on the surroundings. In steps 3 and 4, the surroundings do work on the ideal gas. The amount of work done by the ideal gas on the surroundings in steps 1 and 2 exceeds the amount of work done on the ideal gas in steps 3 and 4, so the net effect is a positive amount of work done by the ideal gas on the surroundings. Heat flows out of the higher temperature reservoir and into the ideal gas in step 1. During the cycle, some of the energy that flows into the ideal gas is used to do work, and some of it flows out of the gas into the lower temperature reservoir in step 3. In one cycle, none of the energy that flows into the ideal gas in step 1 remains in the ideal gas. This is known because at the end of one cycle, the ideal gas is back in its initial state, meaning, among other things, that it has the same internal energy with which it started.

For a Carnot engine, Q_C is the amount of heat that flows out of the ideal gas in step 3, the isothermal compression of the ideal gas. Also, for the Carnot engine, Q_H is the amount of heat that flows into the ideal gas in step 1, the isothermal expansion of the ideal gas. Therefore:

$$\eta_{carnot} = 1 - \frac{Q_{CD,3}}{Q_{AB,1}}$$

As stated before, given that steps 3 and 1 are isothermal processes, the internal energy of the ideal gas does not change in those steps, so $Q_{CD,3} = W_{CD,3}$ and $Q_{AB,1} = W_{AB,1}$. Hence:

$$\eta_{carnot} = 1 - \frac{W_{CD,3}}{W_{AB,1}}$$

Above we stated that it can be shown that $W_{CD,3} = nRT_3 \ln (V_C/V_D)$ and $W_{AB,1} = nRT_1 \ln (V_B/V_A)$. Hence:

$$\eta_{carnot} = 1 - \frac{nRT_3 \ \ln(V_C/V_D)}{nRT_1 \ \ln(V_B/V_A)}$$
$$\eta_{carnot} = 1 - \frac{T_3 \ \ln(V_C/V_D)}{T_1 \ \ln(V_B/V_A)}$$

For an ideal gas, $\frac{V_C}{V_D} = \frac{V_B}{V_A}$ which means that $\ln\left(\frac{V_C}{V_D}\right) = \ln\left(\frac{V_B}{V_A}\right)$ which means that $\frac{\ln(V_C/V_D)}{\ln(V_B/V_A)} = 1$. Hence:

$$\eta_{carnot} = 1 - \frac{T_3}{T_1}$$

Now, T_3 is the temperature of the ideal gas when it is in contact with the hot reservoir at temperature T_H, and T_1 is the temperature of the ideal gas when it is in contact with the cold reservoir at temperature T_C, so the expression of the efficiency of the Carnot engine in terms of the temperatures of the reservoirs is given in equation (20.6) as,

$$\eta_{carnot} = 1 - \frac{T_C}{T_H}. \tag{20.6}$$

Remember the Carnot engine is the most efficient engine possible and is the one to which all heat engines are compared, so equation (20.6), sets the upper limit of efficiency of any heat engine. This is the theoretical maximum efficiency for a real heat engine operating between two thermal reservoirs, one at a higher temperature, T_H, and one at a lower temperature, T_C. This is the use of the Carnot cycle heat engine, since it is not a practical engine that one could build.

Example

Calculate the highest efficiency possible of a heat engine operating between the boiling point and the freezing point of H_2O at STP.

Solutions

Starting with the Carnot efficiency, equation (20.6), and converting the temperatures from °C to K, gives:

$$\eta_{carnot} = 1 - \frac{T_C}{T_H} = 1 - \frac{(0°C + 273.15)}{(100°C + 273.15)} = 1 - \frac{(273.15\ K)}{(373.15\ K)} = 0.268 = 26.8\%$$

20.4.7 SECOND LAW OF THERMODYNAMICS (FIRST FORM)

This leads to the first form of the **second law of thermodynamics**, which states that the theoretical upper limit on the efficiency of a heat engine operating between two temperatures (T_H and T_L) is not 100%, but instead the Carnot efficiency given by equation (20.6). In other words, the efficiency η of any heat engine must meet the requirement that its efficiency is less than a Carnot engine operating between the same temperature range, expressed in equation (20.7) as

$$\eta \le \eta_{Carnot}. \tag{20.7}$$

20.4.8 REFRIGERATION SYSTEMS

How does a refrigerator or an air conditioner work? Have you ever thought about how a refrigerator cools down the food in its compartment and makes the ice in the freezer? How can an air conditioner blow cold air even on the hottest day?

The answer is that a refrigerator is a type of heat pump, which is fundamentally a heat engine run in reverse. Instead of allowing heat to flow in its natural direction from hot to cold and extracting work from this flow, a heat pump moves heat from cold to hot by doing work on the system. The key components of a refrigerator are the compressor, the throttle valve, and the heat exchangers. The compressor is a device that sounds like its name. It is used to pressurize the refrigerant by pushing down on it like the air in a bike air pump. The compressor has one-way valves that open to let the low-pressure refrigerant in, and after compression, lets the high-pressure refrigerant out into the system. Note, all valves are denoted with an x in Figure 20.9, which is a diagram of the compressor and throttle valves.

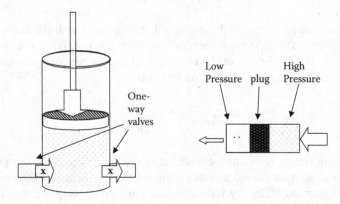

FIGURE 20.9 Compressor is comprised of a piston-cylinder arrangement with an input and output valve (**X**). Throttle valve is simply a porous plug (like a cork) that separates a high-pressure from a low-pressure side.

The throttle valve keeps the high-pressure refrigerant separate from the low-pressure refrigerant. A simple throttle valve is a cork jammed into a pipe. No matter how hard you push one gas on the high-pressure side, the cork will only let a little gas through to the other side of the pipe; that is, only a tiny number of molecules will make it through the cork per second. This effectively separates the high-pressure fluid pushing on the cork from the low-pressure fluid leaking through the cork plug.

In practice, a throttle valve can simply be a blockage in the pipe with a tiny hole drilled through it. A schematic diagram of a refrigerator is given in Figure 20.10.

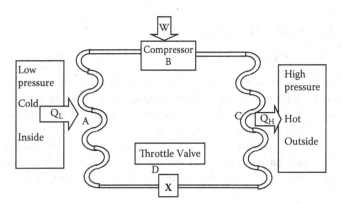

FIGURE 20.10   Schematic diagram of a refrigerator.

The heat exchangers are simply coils of metal tubing filled with a refrigerant. This refrigerant is a fluid that has a high specific heat and goes through phase changes in the temperature region of the device's operation. One of the most infamous refrigerants was Freon, which is a very stable Chlorofluorocarbon. It has excellent properties, but the down side is that it created a hole in the ozone layer. It has since been replaced with other chemicals, which are hopefully less detrimental to our planet.

Since this is a cycle, we can start our analysis at any point. So, our explanation starts at point A, with a cold refrigerant fluid in the heat exchanger, which is in thermal contact with the food in the compartment. As the refrigerant moves through the (Cold) heat exchanger it absorbs energy from the food and warms and changes from a fluid to a gas (evaporates). The refrigerant is then compressed, B, to a high pressure and thus a high temperature. It then moves into the (Hot) heat exchanger, where it exchanges heat to the outside world at point C. The refrigerant, still at a high pressure but cooler, is pushed through the throttle valve and into the low-pressure side of the system. The process starts all over again. For an air conditioner, a fan blows across the cold heat exchanger, and the hot heat exchanger is outside the house.

20.5   ANSWER TO CHAPTER QUESTION

Even though the human metabolism does not convert the food we eat or the beverages we drink into heat to power our body, we can still use the lessons of the heat engine to understand the human body's need to release heat. We ingest fuel in the form of food and drink and convert the chemical composition of the food into usable energy through chemical processes. The process by which we do this is called our metabolism. If we are just sitting around, our metabolism is still running, converting fuel, operating our vital systems, and building cells. It is estimated that the human metabolism is about 20% efficient, so there is a lot of wasted energy that needs to be exhausted. So, even when we are sitting around our bodies need to get rid of the excess energy released from the metabolism process, since our bodies, like all engines, are not 100% efficient. We exhaust the excess energy through radiation in the infrared (wavelength of 9350 nm or around 9 microns), conduction between our skin and the surrounding air, through the heated water we exhale in our breath, and through evaporation of perspiration from our skin. Thus, when it is hotter than 98.6°F, many of these processes don't work very well and our bodies cannot exhaust the extra heat. So, if the average human metabolic rate if we are just sitting around is about 100 W, if we do work, like walking, lifting objects, or running, the metabolic rate increases to 200 W, 300 W, or even 500 W. So, we need to get rid of even more heat. A cooler temperature of 70°F allows our bodies to expel the extra heat efficiently, and thus we feel comfortable.

20.6   QUESTIONS AND PROBLEMS

20.6.1   MULTIPLE CHOICE QUESTIONS

Questions 1–2: The following is a description of a thermodynamic process of a fixed amount of ideal gas in a piston-cylinder arrangement that is graphed in the PV-diagram in Figure 20.11. A gas can be taken from A to C by two different processes, A to B and then to C or directly from A to C, as shown in the diagram. During the direct process A to C, 20 J of work are done by the system and 30 J of heat are added to the system. During the process A to B to C, 25 J of heat are added to the system. Note that each point on the graph corresponds to one equilibrium state of the system with one value of pressure, one value of volume, one temperature, and one internal energy.

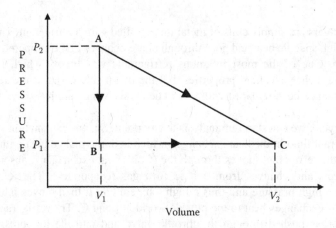

FIGURE 20.11   PV-diagram for multiple choice questions 1–2.

1. What is the change in internal energy of the gas in going directly from A to C?
   A. 5 J
   B. 10 J
   C. 15 J
   D. 20 J
   E. 25 J
2. How much work is done by the system during the processes from A to B to C?
   A. 5 J
   B. 10 J
   C. 15 J
   D. 20 J
   E. 25 J

Questions 3–8: The following is a description of a thermodynamic process of a fixed amount of ideal gas in a piston-cylinder arrangement that is graphed in the PV-diagram in Figure 20.12.

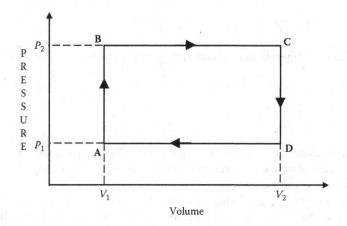

FIGURE 20.12   PV-diagram for multiple-choice questions 3–8.

A gas is contained in piston-cylinder arrangement, and it goes from (being in the state corresponding to) A, to B, to C, to D, and back to A, as depicted in the diagram. The internal energy of the gas at points A, B, C, and D, is 10 J, 20 J, 50 J, and 30 J, respectively. During the processes from (B to C) and from (D to A), 50 J of heat flows into the gas, and 30 J of heat is removed from the gas, respectively.

3. What is the heat input into the gas in the process from A to B?
   A. 0 J
   B. 10 J
   C. 20 J
   D. 30 J
   E. 40 J
4. How much work is done by the gas during the processes from B to C?
   A. 0 J
   B. 10 J
   C. 20 J
   D. 30 J
   E. 40 J
5. How much work is done on the gas during the processes from C to D?
   A. 0 J
   B. 10 J
   C. 20 J
   D. 30 J
   E. 40 J
6. What is the heat output of the gas in the process from C to D?
   A. 0 J
   B. 10 J
   C. 20 J
   D. 30 J
   E. 40 J

7. How much work is done on the gas during the processes from D to A?
   A. 0 J
   B. 10 J
   C. 20 J
   D. 30 J
   E. 40 J

8. What is the net change in the internal energy of the gas for one complete cycle?
   A. 0 J
   B. 10 J
   C. 20 J
   D. 30 J
   E. 40 J

9. In the Carnot cycle, an ideal gas undergoes two adiabatic processes, an adiabatic expansion and an adiabatic compression. Which one of the following statements about those processes is most correct?
   A. Heat flows out of the gas during the compression and into the gas during the expansion.
   B. Heat flow into the gas during the compression and out of the gas during the expansion.
   C. Heat flows into the gas during both processes.
   D. Heat flows out of the gas during both processes.
   E. No heat flows into or out of the gas during either process.

10. In the Carnot heat engine cycle, heat flows into the ideal gas (the working fluid of a Carnot engine) at one temperature and flows out of the ideal gas at another temperature. At which temperature does which heat flow occur?
    A. Heat flows into the ideal gas at the higher temperature, and heat flows out of the ideal gas at the lower temperature.
    B. Heat flows out of the ideal gas at the higher temperature, and heat flows into the ideal gas at the lower temperature.

20.6.2 Problems

1. A piston-cylinder arrangement is filled with 0.03 kg of a gas, which has a specific heat at constant pressure of 2 J/(kg °C). The gas, initially at 20°C, fills an initial volume of 0.25 cm^3 and is at atmospheric pressure. The system is heated and allowed to expand at constant pressure to twice its initial volume, at which point it has a temperature of 50°C. Assuming the atmospheric pressure is 100 kPa, find, for this process, the:
   a. Heat input into the gas.
   b. Work done by the gas.
   c. Change in the internal energy of the gas.

2. What would the efficiency of a Carnot heat engine be if it could be operated between a hot thermal reservoir at a temperature of 5°C, and a cold thermal reservoir at 0 K? (Note that there is no such thing as a thermal reservoir at 0 K.)

3. What percentage of the heat taken in by a heat engine that has half the theoretical maximum possible efficiency and is operating between a thermal reservoir of temperature 20°C and a thermal reservoir of temperature 650°C, has to be dumped to the lower temperature thermal reservoir?

4. An ideal gas undergoes a process described in the PV-diagram in Figure 20.13 The gas starts at the pressure and volume of point **A**, compresses to **B**, expands to **C**, and then its pressure increases until it is back at **A**.

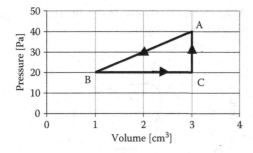

FIGURE 20.13 PV-Diagram for problem 4.

Calculate the work done by the ideal gas for the complete cycle **ABCA**.

5. A piston-cylinder, like the one in Figure 20.1, is fitted with a circular piston with an area of 0.02 m² and filled with 0.004 m³ of an ideal gas. A pressure sensor records the pressure of the ideal gas as 100,500 Pa on a day when the atmospheric pressure is 100,000 Pa. When 1 kJ of heat is added to the ideal gas, the piston rises 20 cm (0.2 m) at a constant pressure. What is the change in internal energy of the ideal gas in the piston-cylinder arrangement during this process?

6. A cylinder is fitted with a movable piston, which has a mass of 8 kg and a cross sectional area of 60 cm². The cylinder is filled with a volume of an ideal gas so that the piston is resting on a 30 cm column of an ideal gas on a day in which the atmospheric pressure is 100 k Pa. At the initial equilibrium position, the gas is at a temperature of 30°C. Then, 400 J of heat is added to the gas so that the gas temperature increases and the piston and block rises 20 cm to a new equilibrium position. Compute the change in internal energy of the gas during the heating processes in which the piston and block rise 20 cm.

7. & 8. A piston-cylinder arrangement, like the one in Figure 20.1, has a piston with a cross-sectional area of 10 cm². Three-tenths (0.3) of a mole of an ideal gas is enclosed in the cylinder. The PV-diagram of a four step processes done with this piston-cylinder is given in Figure 20.14. Note: 1 kPa = 1000 Pa.

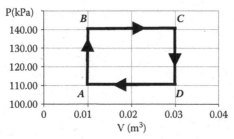

FIGURE 20.14 PV-Diagram for problems 7 & 8.

7. Compute the net work done by the gas in one complete cycle.

8. If 3,922 J of heat is added to the gas to achieve the expansion of B to C, compute the change in internal energy of the gas from B to C.

9. A piston-cylinder arrangement like the one in Figure 20.1 is fitted with a movable piston that has a mass (*m*) and a cross-sectional area of 1.75 m². The cylinder is filled with a volume of an ideal gas so that the piston is resting on a column of an ideal gas that has a height of 12 cm. When 30 kJ of heat is added to the ideal gas in the cylinder, the piston rises 8 cm from with a constant pressure of 102 k Pa for the ideal gas in the cylinder. Compute the change in internal energy of the gas during the heating processes in which the piston and block rise 8 cm.

10. Doctors generate a PV-diagram, like the one in Figure 20.15, of a patient's heart to better understand the performance of the organ.

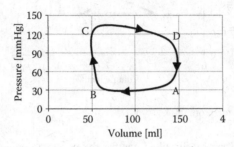

FIGURE 20.15   PV-diagram for problem 10.

In the PV-diagram, the diastolic ventricular filling occurs from A to B, the isovolumetric contraction occurs from B to C, the systolic ejection occurs from C to D, and isovolumetric relaxation from D to A. Given that 1 mmHg is equal to 133.32 Pa and 1 ml is equal to $1 \times 10^{-6}\,m^3$, compute the work per cycle of the heart from which Figure 20.15 was produced. Do your best to approximate the area enclosed.

21 Thermodynamics – Entropy

21.1 INTRODUCTION

Entropy is the topic for this chapter. Starting with the form of the second law of thermodynamics derived in the previous chapter, the change in entropy is established as an important quantity. Through several examples in which the change in entropy is computed, this often-misunderstood quantity is demystified through calculations associated with changes in temperature and/or the changes in state of an object. A statistical representation of entropy is given, and through several statistical examples, entropy and the second law of thermodynamics is explained as a consequence of probability.

21.2 CHAPTER QUESTION

Does the evolution of life on earth violate the second law of thermodynamics? This question is answered at the end of this chapter after the concepts of entropy and the second law of thermodynamics are established.

21.3 DEFINITION OF ENTROPY AND THE CARNOT CYCLE

In Chapter 20, the first version of the second law of thermodynamics was introduced through the study of the Carnot heat engine. This first form of the second law states that the maximum possible efficiency for a heat engine operating between T_C and T_H is that of the Carnot engine. It is summarized in equation (20.6) as $\eta \leq \eta_{\text{Carnot}}$, which states that the efficiency η of any heat engine must have an efficiency that is less than or equal to that of a Carnot engine. In Chapter 20, it was also established that the efficiency of any heat engine can be expressed as $\eta = 1 - \frac{Q_C}{Q_H}$ and that the efficiency of a Carnot engine can be expressed as $\eta_{\text{carnot}} = 1 - \frac{T_C}{T_H}$. Hence, equation (20.6) ($\eta \leq \eta_{\text{Carnot}}$) can be written as:

$$1 - \frac{Q_C}{Q_H} \leq 1 - \frac{T_C}{T_H}$$

Adding $\frac{Q_C}{Q_H} + \frac{T_C}{T_H} - 1$ to both sides yields: $\frac{T_C}{T_H} \leq \frac{Q_C}{Q_H}$

Multiplying both sides by $\frac{Q_H}{T_C}$ and switching sides results in equation (21.1) as,

$$\frac{Q_C}{T_C} \geq \frac{Q_H}{T_H} \tag{21.1}$$

This leads us to the definition of another state variable, **entropy,** which is represented by the symbol S. The quantity on either side of equation (21.1) is defined as the change in entropy so that this can be expressed in equation (21.2) as,

$$\Delta S = \frac{Q}{T} \tag{21.2}$$

DOI: 10.1201/9781003308072-22

So, equation (21.1) can be combined with equation (21.2) to generate the second form of the second law of thermodynamics, which can be written as the following equation,

$$\Delta S_C \geq \Delta S_H$$

or more commonly as equation (21.3):

$$\Delta S \geq 0, \text{ with } \Delta S = (\Delta S_C - \Delta S_H) \tag{21.3}$$

This form of the **second law of thermodynamics** states that the entropy of a system, including the heat engine and the surroundings, always increases during a natural process. This is different than most quantities, which can increase or decrease depending on the situation.

This can be understood by recognizing that this form of the second law comes from an investigation of the efficiency of a heat engine. So, entropy is the measure of the amount of thermal energy that cannot be used to do work. When energy is distributed throughout the molecules of a substance as heat, there is something fundamental that does not allow all the energy to be used for the process intended, and there is always energy lost. This leads to the concept of the irreversible processes of a real heat engine. If heat is put into a system, all the heat cannot be used to do work, because the entropy of the process must increase. Thus, entropy measures the irreversible changes of a system. Remember, entropy is defined such that, when an amount of heat Q flows into a system at a constant temperature T, the entropy of the system increases by an amount that is equal to Q/T, *and when* heat Q flows out of a system at a constant temperature T, the entropy of the system decreases by an amount that is equal to $-Q/T$.

The first application of entropy is to study the working fluid, normally an ideal gas, in a heat engine. For any complete cycle, the working fluid returns to the initial state of the cycle. That means, at the end of each cycle, the entropy of the working fluid is the same as what it was at the start of each cycle. Thus, the change in the entropy of the working fluid in one cycle is zero. This is true for any heat engine, not just a Carnot engine. Referring back to Figure 20.5 of the flow diagram of a heat engine, the second law requires that the change of entropy of the thermal reservoirs also be considered. For this system, the entropy change of a thermal reservoir is just the quotient of the heat flowing into or out of it divided by the temperature of the reservoir, because a thermal reservoir stays at one constant temperature. In the course of one cycle, the entropy of the hotter thermal reservoir decreases by Q_H/T_H because an amount of heat Q_H flows out of it while it is at temperature T_H. Also, during the course of one cycle, the entropy of the colder thermal reservoir increases by Q_C/T_C because an amount of heat Q_C flows into it while it is at a temperature T_C. So, the total change in entropy of the working fluid and the surroundings is:

$$\Delta S = \Delta S_{\text{working fluid}} + \Delta S_{\text{cold reservoir}} + \Delta S_{\text{hot reservoir}}$$
$$\Delta S = 0 + Q_C/T_C - Q_H/T_H$$
$$\Delta S = Q_C/T_C - Q_H/T_H$$

Equation (21.1) states that $Q_C/T_C \geq Q_H/T_H$, which means that

$$\Delta S \geq 0$$

That is, the total entropy change of a system plus its surroundings is always ≥ 0. This is the second law of thermodynamics, and it is true not just for heat engines but for all thermodynamic systems. Note that the "equal to" part of the "greater than or equal to sign" applies in the case of a Carnot engine.

The fact that, during any process, the total change of the entropy of a system plus its surroundings is greater than zero, does not mean that the entropy of every system during every process can only increase. In the case of the heat engine discussed above, the entropy of the hotter thermal reservoir actually decreased. The entropy of the working fluid of the heat engine increased and decreased during a cycle for a net change of zero for the cycle. Only the entropy of the colder thermal reservoir increased, but the total change in the entropy of all three items taken together confirmed the second law so that $\Delta S_{total} \geq 0$.

Energy obeys a conservation law that can be stated as: Energy can neither be created nor destroyed. Entropy has a different type of law, in that it can be created but it can't be destroyed.

Here is a simple example, illustrated in Figure 21.1.

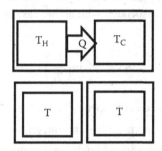

FIGURE 21.1 A hot and cold container in thermal contact, then separated.

Suppose a hot thermal reservoir, at a temperature of T_H, is put into an insulated container with a colder thermal reservoir, at a temperature of T_C. Assuming the thermal reservoirs are in thermal contact with each other and insulated from the surrounding, an amount of heat, Q, flows from the hotter thermal reservoir to the colder thermal reservoir. Thus, the entropy of the hotter reservoir decreases by the amount Q/T_H and the entropy of the colder reservoir increases by Q/T_C for a net change in the entropy of the pair of thermal reservoirs of

$$\Delta S = Q/T_C - Q/T_H$$

Since $T_C < T_H$ are both in the denominator of their respective terms, the first term is greater than the second term, and the net change in the entropy of the pair of thermal reservoirs is greater than zero. The change in entropy is positive, so entropy has been created. In any actual thermodynamic process, the total entropy of a system plus its surroundings increases. Consider the simple process of spontaneous heat flow from a hot object to a cold object described in a previous section. The total entropy of the pair of objects increases; therefore, this system is considered *irreversible*.

21.3.1 ENTROPY EXAMPLES

Example 1

Compute the change in entropy of 2 kg of H_2O when transformed from liquid water at 100°C at atmospheric pressure to H_2O gas at 100°C at atmospheric pressure.

$$Q = mL_v = (2 \text{ kg}) \, 2260000 \text{ J/kg} = 14520000 \text{ J}$$
$$T = 100°C = 373.15 \text{ K}$$
$$\Delta S = Q/T = 4520000 \text{ J}/(373.15 \text{ K}) = 12113 \text{ J/K}$$

Example 2

Calculate the change in entropy of 2 kg of ice when it melts at 0°C.

$$\Delta Q = mL_m = (2\text{kg})334000 \text{ J/kg} = 668000 \text{ J}$$
$$T = 0°C = 273.15 \text{ K}$$
$$\Delta S = \Delta Q/T = 668000 \text{ J}/(273.15 \text{ K}) = 2445.5 \text{ J/K}$$

Thus far, the temperature of the objects for which the entropy has been computed has been kept at a constant value, so the change in entropy was computed with equation (21.2). If the temperature of the object is changing as heat is flowing into or out of the system, the calculation of entropy must be computed with a different expression. In the case of an object that is undergoing a change of temperature, the change in entropy associated with an increase in temperature is given in equation (21.4) as

$$\Delta S = mc \ln\left(\frac{T_f}{T_i}\right), \tag{21.4}$$

where, for the object, m is the mass, c is the specific heat, T_f is the final temperature, and T_i is the initial temperature. This expression comes from the integral of $\Delta S = \int mc\frac{dT}{T} = mc \ln\left(\frac{T_f}{T_i}\right)$, but it can be used in solutions of problems without applying concepts from calculus.

Example 3

A 200 g block of iron at 100°C is dropped into an insulated container filled with 500 g of liquid water at 20°C.

  A. What is the final temperature of the water and iron block?
     See the specific heat example in Chapter 19, in which a final temperature was found:
     $T_f = 23.3°C$
        The heat for each of the temperature changes of water and iron follows:

$$Q_{H2O} = m_{H2O}\, c_{H2O}(T_f - T_{iH2O}) = (0.5 \text{ kg})\, (4,186 \text{ J/kg°C})\, (23.3°C - 20°C) = 6900 \text{ J}$$
$$Q_{Fe} = m_{Fe}\, c_{Fe}(T_f - T_{iFe}) = (0.2 \text{ kg})\, (450 \text{ J/kg°C})\, (23.3°C - 100°C) = -6900 \text{ J}$$

  These magnitudes are equal to two significant digits.

  B. What is the change of entropy of the water and iron block?

Since there is a change in temperature in this process, the change in entropy needs to be computed with equation (21.4).
  Remember that the temperatures must be in kelvin for entropy calculations, and since the temperature is changing in this process, the change in entropy must be calculated with the

$$\Delta S_{H2O} = m_{H2O}\, c_{H2O} \ln\left(\frac{T_f}{T_{iH2O}}\right) = (0.5 \text{ kg})\left(4,186\frac{J}{\text{kg K}}\right)\left(\ln\left(\frac{296.3 \text{ K}}{293 \text{ K}}\right)\right) = 23.4\frac{J}{K}$$

$$\Delta S_{Fe} = m_{Fe}\, c_{Fe} \ln\left(\frac{T_f}{T_{iFe}}\right) = (0.2 \text{ kg})\left(450\frac{J}{\text{kg K}}\right)\left(\ln\left(\frac{296.3 \text{ K}}{373 \text{ K}}\right)\right) = -20.7\frac{J}{K}$$

So, the net change in entropy is $\Delta S = \Delta S_{H2O} + \Delta S_{Fe} = 23.4$ J/°K + -20.7 J/°K = 2.7 J/°K. The entropy increases by about 12% during the process.

21.4 STATISTICAL MECHANICS & ENTROPY

The second law of thermodynamics describes what happens to the change of entropy (ΔS), but it doesn't explain what entropy is and why entropy is often explained as a measure of disorder. To do this, the concepts of statistical mechanics of the microscopic world of atoms and molecules must be employed. The starting points of understanding this analysis are the concepts of the groupings used to label the grouping of atoms and molecules. These groupings are called **microstates**, which are the possible arrangements that achieve the arrangement that satisfies the characteristic of the group called a **macrostate**.

21.4.1 Definition of Multiplicity and Entropy

Consider a simple system consisting of four distinguishable particles (P_1, P_2, P_3, P_4), each of which can only have a non-negative integer number of joules of energy (0 J, 1 J, 2 J, …). Further suppose that the system is in a macrostate for which the total energy is 2 joules (2 J). The possible microstates for this (2 J) macrostate are given in Table 21.1.

TABLE 21.1

Microstates for a Four-Particle System with 2 J of Energy

Particle ➡	P_1	P_2	P_3	P_4
Energy per particle	2	0	0	0
Energy per particle	0	2	0	0
Energy per particle	0	0	2	0
Energy per particle	0	0	0	2
Energy per particle	1	1	0	0
Energy per particle	1	0	1	0
Energy per particle	1	0	0	1
Energy per particle	0	1	1	0
Energy per particle	0	1	0	1
Energy per particle	0	0	1	1

So, the total number of microstates, the multiplicity Ω, consistent with a macrostate of a total energy of 2 J is:

$$\Omega = 10$$

Entropy is defined in terms of the multiplicity Ω of a macrostate, as given in equation (21.5) as,

$$S = k_B \ln(\Omega), \tag{21.5}$$

where k_B is still the Boltzmann constant, which has a value of 1.38065×10^{-23} J/K. So, for the example of the four-particle system with a total energy of 2 J, the entropy of the 2 J macrostate is:

$$S_{2J} = k_B \ln(\Omega) = (1.38 \times 10^{-23} \text{ J/K}) \ln(10) = 3.178 \times 10^{-23} \text{ J/K}.$$

The definition of entropy given in equation (21.5) above is consistent with the classical thermodynamics statement that entropy is a state variable whose value increases by an amount Q/T when an amount of heat Q flows into the system while the system is at a constant temperature T. The simple example above can be used to show how adding energy to a system increases the entropy of that system.

For the case of the system of four particles discussed above, suppose some energy flows into the system so the internal energy increases from 2 J to 3 J. The energy can be divided up as shown in Table 21.2.

TABLE 21.2
Microstates for a Four-Particle System with 3 J of Energy

Particle ➡	P_1	P_2	P_3	P_4
Energy per particle	3	0	0	0
Energy per particle	0	3	0	0
Energy per particle	0	0	3	0
Energy per particle	0	0	0	3
Energy per particle	2	1	0	0
Energy per particle	2	0	1	0
Energy per particle	2	0	0	1
Energy per particle	0	2	1	0
Energy per particle	0	2	0	1
Energy per particle	1	2	0	0
Energy per particle	0	0	2	1
Energy per particle	1	0	2	0
Energy per particle	0	1	2	0
Energy per particle	1	0	0	2
Energy per particle	0	1	0	2
Energy per particle	0	0	1	2
Energy per particle	1	1	1	0
Energy per particle	1	1	0	1
Energy per particle	1	0	1	0
Energy per particle	0	1	1	1

This makes for a total of $\Omega = 20$ microstates. Hence, the new value of entropy is

$$S_{3J} = k_B \ln(\Omega) = (1.38 \times 10^{-23} \text{ J/K}) \ln(20) = 4.134 \times 10^{-23} \text{ J/K}.$$

So, it is clear that as energy is added to a system of four particles, the energy increases from 2 J to 3 J and the entropy of the system increases from 3.178×10^{-23} J/K to 4.134×10^{-23} J/K, so the expression of entropy in equation (21.2), $\Delta S = \frac{Q}{T}$, holds for this system.

21.4.2 MULTIPLICITY AND ENTROPY IN A MANY PARTICLE SYSTEM

A real thermodynamic system is comprised of many particles that have energy, over a wide range of energies. In addition, the greater the total energy of the system, the greater the multiplicity, and hence, the greater the entropy. This is consistent with the idea that when energy flows into a system, the entropy of the system increases. For a simple model of solids made up of a collection of vibrating atoms, the multiplicity for each of the microstates is given in equation (21.6) as

$$\Omega = \frac{(q + N - 1)!}{q!(N - 1)!},\tag{21.6}$$

where q is the energy in a specific solid, N is the number of oscillating atoms in the object, and the (!) represents the factorial of a number. For example, $5! = 5 * 4 * 3 * 2 * 1 = 120$. Equation (21.6) is a fundamental mathematical function in combinatorics, which is the study of ways in which different systems can be combined.

For a simple example depicted in Figure 21.2, a system consists of two objects that are in thermal contact with each other, but insulated from the rest of the universe.

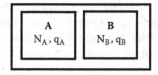

FIGURE 21.2 A hot and cold container in thermal contact, then separated.

Given a total energy of $q = 7$ separate units of energy, one object made up of $N_A = 4$ particles and the other made up of $N_B = 3$ particles, the multiplicity, probability, and entropy can be computed and graphed for each particle as a function of the energy in solid A in Figure 21.2. The energy in solid B is just the total energy minus the energy in A. The graphs of the multiplicities of each solid, A and B, are given in Figures 21.3 and 21.4.

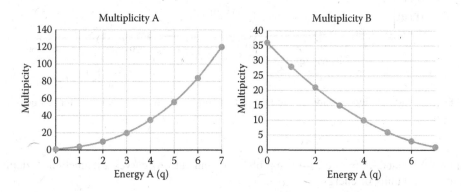

FIGURE 21.3 Multiplicity of each solid, A and B, as a function of the energy in object A.

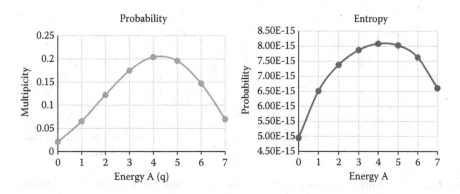

FIGURE 21.4 Multiplicity and entropy of the combination of both solids, A and B, as a function of the energy in object A, with $q = 7$, $N_A = 4$, and $N_B = 3$.

This is done using the expression of the multiplicity given in equation (21.6) and distributing the 7 units of energy among the two particles.

The total multiplicity is computed as a product of the two multiplicities $((\Omega_A)(\Omega_B))$, and the probability is computed by dividing the total multiplicity by the total number of microstates for this system, which is 1,726. The entropy is then calculated with equation (21.5), and the total multiplicity as $S = k_B \ln((\Omega A)(\Omega B))$.

Notice, for this system the maximum of the probability and the maximum of the entropy graphs coincide at $q = 4$. This is the most probable arrangement of the energy for this system, with 4 units of energy in A and 3 units of energy in B. This is the equilibrium energy distribution for this system.

If the number of particles in each solid and the amount of energy in the system is increased, it is interesting to notice how the multiplicity, probability, and entropy change. For example, if the energy is increased to $q = 60$ units, solid A is increased to 10 particles ($N_A = 10$), and solid B to 20 particles ($N_B = 20$), the individual multiplicities of each object look similar to those in Figure 21.3, with larger numbers and steeper slopes, as shown in Figure 21.5.

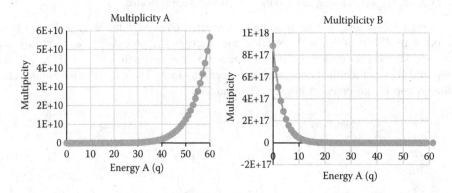

FIGURE 21.5 Multiplicity of solids, A and B, as a function of the energy in object A for $q = 60$, $N_A = 10$, and $N_B = 20$.

On the other hand, in Figure 21.6, the probability and entropy curves are much more defined at the center of 20 units of energy.

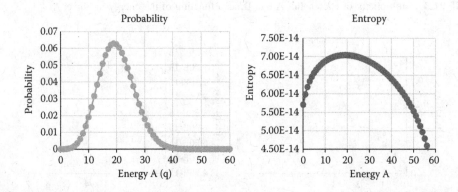

FIGURE 21.6 Probability and entropy of the combination of both solids, A and B, as a function of the energy in object A for $q = 60$, $N_A = 10$, and $N_B = 20$.

So, again the distribution of energy matches the distribution of particles ($N_A = 10$, $N_B = 20$, so $q_A = 20$ and $q_B = 40$). That is, the most probable distribution of energy (q) among the two solids can be expressed by the ratio in equation (21.7) as

$$\frac{N_A}{N_B} = \frac{q_A}{q_B}. \tag{21.7}$$

Also, as the number of particles and the amount of energy is increased, the probability curve narrows around the value found by equation (21.7).

Again, increasing the number of particles in each solid to solid A containing 400 particles ($N_A = 400$) and solid B made up of 1,200 particles ($N_B = 1200$) and a total energy of $q = 100$ units does not change the structure of individual multiplicities, but it does continue to narrow the probability and entropy curves (see Figure 21.7).

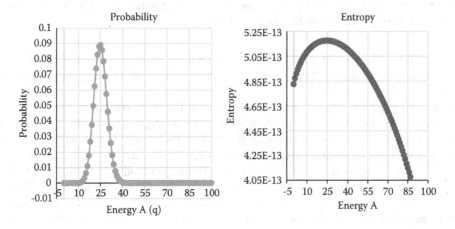

FIGURE 21.7 Probability and entropy of the combination of both solids, A and B, as a function of the energy in object A for $q = 100$, $N_A = 400$, and $N_B = 1200$.

The probability and entropy graphs both peak at 25 units of energy, so the distribution of energy matches the distribution of particles ($N_A = 400$, $N_B = 1200$, so $q_A = 25$ and $q_B = 75$). That is, $\frac{N_A}{N_B} = \frac{q_A}{q_B}$ so $\frac{400}{1200} = \frac{25}{75}$. The total number of the possible microstates for this system is a staggering 5.7356×10^{163}.

One last increase, this time increasing the number of particles in each solid and the amount of energy in the system. The total energy of the system is set at $q = 200$ units, the number of particles in solids A and B are increased to 250 and 1,000 particles, respectively. Therefore, using equation (21.7), the maximum probability and maximum entropy should occur at a ratio of $\frac{N_A}{N_B} = \frac{250}{1000} = \frac{1}{4}$ so $\frac{q_A}{q_B} = \frac{1}{4} = \frac{40}{160}$. So, the maximum probability should be at a value of $q_A = 40$. The total number of the possible microstates for this system is a staggering 1.1431×10^{251}. The probability and entropy graphs for the combination described in this paragraph are given in Figure 21.8.

Remember that there are Avogadro's number of atoms in a mole of a substance, so that means that there are approximately 6.022×10^{23} aluminum atoms, which is about 60 regular-sized paperclips. That is a lot of atoms in a small amount of mass. So, as the numbers increase, the probability function gets narrower and narrower, so that there is only one arrangement that has a probability that will occur. These calculations cannot be done on a spreadsheet, and statistical

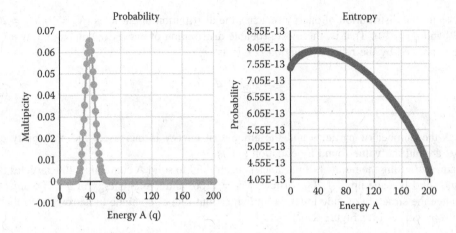

FIGURE 21.8 Probability and entropy of the combination of both solids, A and B, as a function of the energy in object A for $q = 200$, $N_A = 250$, and $N_B = 1000$.

mechanics is full of interesting techniques to compute these multiplicities, probabilities, and entropies for systems with large numbers of particles like a piston-cylinder arrangement filled with an ideal gas.

This analysis provides an explanation of the **second law of thermodynamics**, which states that the entropy of the entire system always increases during a natural process. The graphs clearly point out that the arrangement with the highest probability is one with the maximum entropy. It is a matter of statistics that leads to an increase in entropy. This analysis also provides a reason why thermodynamic reactions only happen naturally in one direction. If a hot object is dropped into a container filled with cold water, the object will cool and the water will warm. The result that the hot object will take some energy from the cold water and get hotter, will never happen-not because it cannot, but because there is so small a probability of it happening that it never does. It turns out that the second law of thermodynamics indicates a direction in which a process can occur, so it is an arrow of time. This type of analysis can be done for all types of systems and is something done in courses like thermodynamics and physical chemistry.

21.4.3 IMPORTANT APPLICATIONS

So, the next time you stir in some cold milk in a cup of hot coffee and the mixture becomes a bit cooler, you will have a new way to explain why the coffee doesn't get hotter and the milk doesn't get colder. The second law of increasing entropy gives a physical reason for this. This statistical argument is known as an arrow of time and is one of the few physical theories that indicates the flow of time in one direction from past to present and not the other way. This idea is used to explain the folding of proteins, as shown in Figure 21.9.

This folding is a critical step in many biological processes in the body. Only when the protein is in the proper folded arrangement can the protein be used by the organism to complete a critical process. A decrease in the number of proteins folded in the correct way in the body may result in diseases such as cancer and Alzheimer's. One way to understand why these proteins fold in on themselves is to use the second law of thermodynamics. In the case of the unfolded protein, the non-polar amino acid residues (H) along the protein are exposed to the water surrounding the protein. This lowers the entropy of the water by restricting the way in which the water molecules

Unfolded Protein   Folded Protein

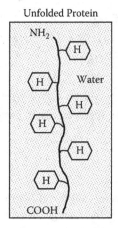 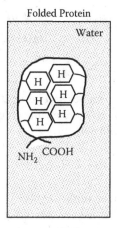

FIGURE 21.9   Folding of proteins.

can be arranged in the area of the protein. When the protein folds in upon itself so that the non-polar amino acids are separated from the water, the entropy of the water increases. This increase in the entropy of the system is the explanation of why proteins fold in upon themselves. This process of folding is known as the hydrophobic effect.

21.5   ANSWER TO CHAPTER QUESTION

It is true that life in general and the evolution of life requires negative entropy. In fact, this is one of the defining characteristics of life. Living organisms themselves use energy to continually organize systems from the cellular to the organismal level. The error that is made by those that use the second law of thermodynamics as a way to argue against evolution is that "although the entropy of the universe must increase with time, the entropy of any part of the universe can decrease with time, as long as the decrease is compensated by an even larger increase in entropy in some other part of the universe".[1]

The following argument was paraphrased from the 2008 *American Journal of Physics* article "Entropy and Evolution" by Daniel Styer.[1]

For life on earth, the most important source of energy is the heating of the earth by the sun. On average, the earth absorbs 4.74×10^{17} J of energy each second from the sun. The average temperature of the earth's surface is about 14.85°C, which is 288 K. So, the net entropy increases on the earth due to solar radiation is:

$$\Delta S_{\text{Earth from Sun}} = Q/T = [(4.74 \times 10^{17} \text{ J})/(288 \text{ K})] = +1.65 \times 10^{15} \text{ J/K}$$

The change in entropy due to the evolution of living organisms must be computed from statistical analysis, since evolution decreases the number of ways organisms can accomplish the task of living. Assuming that because of evolution, each individual organism is 1,000 times more probable than the corresponding individual was 100 years ago, then for every final microstate of $\Omega_f = 1$ there were $\Omega_i = 1000$.

So, compute the change in entropy of an organism due to evolution over 100 years:

$$\Delta S_{evolution\_per\_organism} = S_f - Si = (k)ln(\Omega_f) - (k)ln(\Omega_i) = k[ln(\Omega_f) - ln(\Omega_i)]$$

By the mathematical law of natural logs that: $ln(A) - ln(B) = ln(A/B)$

[1] Styer., D., "Entropy and Evolution", *American Journal of Physics,* 76, (11), 2008.

$$\Delta S_{evolution\_per\_organism} = k\left[ln\left(\Omega_f/\Omega_i\right)\right] = (1.38 \times 10^{-23} \, J/K)[ln\,(1/1000)]$$

$$\Delta S_{evolution\_per\_organism} = k\left[ln\left(\Omega_f/\Omega_i\right)\right] = (1.38 \times 10^{-23} \, J/K)[-6.91]$$

$$\Delta S_{evolution\_per\_organism} = (-9.54 \times 10^{-23} J/K)$$

So, the evolution per organism per second is: $\Delta S_{e\_per\_o} = (-3.02 \times 10^{-32} \, J/K)$

With an estimate of 6×10^{32} organisms on earth, the decrease in entropy due to evolution on earth per second is approximately:

$$\Delta S_{evolution} = (-3.02 \times 10^{-32} \, J/K) * (6 \times 10^{32}) = -18.12 \, J/K$$

So, even with an overestimate of the number of species and the effect of evolution, the change in entropy due to evolution is insignificant relative to the change in entropy due to the sun. So, evolution does not violate the second law of thermodynamics.

21.6   QUESTIONS AND PROBLEM

21.6.1   MULTIPLE CHOICE QUESTIONS

Questions 1–3: A glass is filled with ice cubes at 0°C and then filled with cold water at 40°C. At the instant under consideration, some of the ice has melted and the liquid water has cooled down only as far as 35°C. (The liquid will continue to cool, and the ice will continue to melt.)

1. When the ice melts from a solid at 0°C to a liquid at 0°C, is the change in entropy of the ice plus the liquid that was ice positive, negative, or zero?
   A. positive
   B. negative
   C. zero
2. When the liquid water cools from 40°C to 35°C, is the change in entropy of the liquid water that was originally liquid positive, negative, or zero?
   A. positive
   B. negative
   C. zero
3. From the time when the temperature of the originally-liquid water is 40°C to the time it is 35°C, is the net change in entropy of the ice and liquid water, which came from the ice, positive, negative, or zero?
   A. positive
   B. negative
   C. zero
4. For one complete cycle of a heat engine, like the one in Figure 20.5, is the net change in entropy of the working fluid (the ideal gas) positive, negative, or zero?
   A. positive
   B. negative
   C. zero
5. For one complete cycle of a heat engine, like the one in Figure 20.5, is the net change in entropy of the hot reservoir positive, negative, or zero?
   A. positive
   B. negative
   C. zero

6. For one complete cycle of a heat engine, like the one in Figure 20.5, is the net change in entropy of the cold reservoirs positive, negative, or zero?
   A. positive
   B. negative
   C. zero

7. For one complete cycle of a heat engine, like the one in Figure 20.5, is the net change in entropy of the entire system, including the working fluid (the ideal gas) and both the hot and cold reservoirs, positive, negative, or zero?
   A. positive
   B. negative
   C. zero

8. A tiny drop (0.001 kg) of molten copper, at its melting point (1083°C), is dropped into a large pool of water, which is at a uniform temperature of 20°C. While the molten copper solidifies, from a liquid to a solid, the temperature of the pool remains constant at 20°C. Is the change in the entropy of the pool water as the copper solidifies and cools to the temperature of the pool water positive, negative, or zero?
   A. positive
   B. negative
   C. zero

9. The greater the multiplicity of a macrostate is, the greater the entropy of the system that is in that macrostate.
   A. true
   B. false

10. Consider two systems, system A and system B, side by side, but isolated from each other. If system A has multiplicity Ω_A and system B has multiplicity Ω_B, then the multiplicity of the combination system consisting of both system A and system B is the product $\Omega_A \Omega_B$ since for each one of the Ω_A microstates that system A can be in, system B can be in Ω_B microstates. Given that system A is in a macrostate having entropy S_A and system B is in a macrostate having entropy S_B, what is the entropy of the combination system consisting of both system A and system B (still isolated from each other)?
    A. $S_A + S_B$
    B. $S_A S_B$
    C. $\sqrt{S_A S_B}$
    D. $\sqrt{S_A^2 + S_B^2}$

21.6.2 Problems

1. Find the entropy change for a heating pad at 70°C that conducts 150 J of energy into your muscle.

2. Find the change in entropy of 0.05 kg of spilled ethyl alcohol at 20°C that evaporates into the room temperature air at 20°C. Important data: the specific heat of liquid ethanol is $2.4 \frac{kJ}{kg\,K}$, the specific heat of ethanol gas is $1.8 \frac{kJ}{kg\,K}$, and the heat of vaporization of ethanol is $846 \frac{kJ}{kg}$, at a temperature of 78.73°C. Hint: The ethanol needs to get up to the vaporization temp to vaporize, and then it cools back down to room temperature.

3. As you stand outside on a cold day, your 33°C skin loses 150 J of heat in 1 s to the 5°C air around you. Assume that neither you nor the air change temperature significantly during that second.
   a. How much does your entropy change during that second?
   b. How much does the air's entropy change during that second?
   c. What is the net change in entropy between you and the air?

4. A 100 kW steam engine having 20% the efficiency of a Carnot engine does work at the average rate of 100 kW with a hot reservoir at a temperature of 600°C and a cold reservoir at a temperature of 50°C. Calculate:
   a. the efficiency of the heat engine,
   b. the average rate at which heat flows from the hot reservoir to the working fluid,
   c. the average rate at which heat flows from the working fluid to the cold reservoir,
   d. the average rate at which the entropy of the hot thermal reservoir is changing,
   e. the average rate at which the entropy of the cold thermal reservoir is changing, and
   f. the average rate of entropy generation in the system and surroundings.

5. In an experiment designed to test the thermal conductivity of glass, a person boils water in a container with a flat glass top with a block of ice on top of the glass. The person measures the rate at which liquid water that was ice accumulates to determine the rate at which heat is flowing through the glass. The glass is tilted so that the water formed by the melting of the ice flows into a container as soon as it melts, meaning that the water accumulating in the container is at 0°C. Suppose the water is accumulating at the rate of 2.50 cm³/s.
   a. What is the rate of heat flow out of the steam at 100°C, through the glass and into the ice?
   b. At what rate is entropy being generated by the flow of heat out of the steam and into the ice?

6. A tiny drop (0.001 kg) of molten copper, at its melting point (1083°C), is dropped into a large pool of water, which is at a uniform temperature of 20°C. The heat of fusion (melting) for copper is 134 kJ/kg. While the molten copper solidifies, from a liquid to a solid, the temperature of the pool remains constant at 20°C. While the copper solidifies, compute the change in entropy of:
   a. the tiny drop of copper,
   b. the pool of water, and
   c. the entire system of the tiny drop of copper and the pool water.

7. A ring at 35°C is made up of a mass of 20 g of pure platinum, which melts at 1771°C and has a heat of fusion (melting) of 0.100 J/(kg°C). The ring is dropped into molten lava, which is at a constant temperature of 4000°C. For the entire process of the ring heating up and melting, compute the change in entropy of the:
   a. ring,
   b. the molten lava, and
   c. the system made up of the ring and the molten lava.

8. A penny, made of pure copper with a mass of 2.5 g, is at 37°C. The penny is dropped into a large container filled with pure molten copper at its melting point of 1083°C. While the penny melts, from a solid to a liquid, the temperature of the molten copper remains constant at 1083°C. Compute the change in entropy of the entire system of the penny and the vat of molten copper while the penny melts.

9. Enough heat is caused to flow into 2.00 kg of liquid water at 100°C at atmospheric pressure to turn it all into water vapor at 100°C at atmospheric pressure.
   a. How does the entropy of the water vapor at 100°C compare with the entropy of the liquid water at 100°C?
   b. How does the multiplicity of the water vapor at 100°C compare with the multiplicity of the liquid water at 100°C?

10. Consider a system consisting of five distinguishable particles (P_1, P_2, P_3, P_4, P_5), each of which can only have a non-negative integer number of joules of energy (0 J, 1 J, 2 J, ...). Find the multiplicity and entropy for the case in which the system is in a macrostate for which the total energy is 3 J.

22 Electric Circuits

22.1 INTRODUCTION

The study of electrical circuits is another branch of physics in which conserved quantities play a central role in the analysis. In fact, the well-known term of voltage is just potential energy per charge, and the laws that govern the way voltage is distributed across a circuit are based on conservation of energy. That is, the voltage supplied must equal the voltage lost around a circuit. The other quantity that plays a central role in the study of circuits is electrical current, which is just the rate at which charge moves through the wires. Because charge is a fundamental quantity in nature, it cannot be created or destroyed; the amount of charge into one part of a circuit must equal the amount flowing out that same part of the circuit. This concept will become the second fundamental rule by which circuits are analyzed. The quantity that relates the voltage dropped across a part of the circuit, known as a circuit element, and the current that flows through this element is defined in the chapter as the resistance of the element. The focus of this chapter is on the introduction of the quantities employed to study electrical circuits, the laws that govern those quantities, and some of the techniques applied when analyzing electrical circuits.

22.2 CHAPTER QUESTION

In your home, as you turn on more lights and other appliances, like your TV or computer, they all come on and don't seem to affect the other devices that are already operating. How does this work? Is there a sensor that increases the feed of electricity to the house as you turn on more stuff or is it set up so that it just happens? This question will be answered at the end of the chapter using the concepts of current, voltage, and resistance defined in this chapter.

22.3 INTRODUCTION TO CIRCUIT ANALYSIS

22.3.1 VOLTAGE (V)

The starting place for understanding electricity is the definition of a familiar word, voltage (**V**). This is a word that grew from the term *electric potential*, which is the electric potential energy per unit change in a region of space.

Consider a region of space in which there exists an electric field (**E**), represented by the arrows pointing from the positive line of charges on the right to the negative line of charges on the left in Figure 22.1.

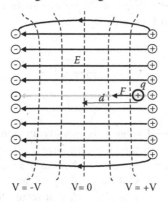

FIGURE 22.1 Electric field of two lines of opposite charge and a test charge q.

DOI: 10.1201/9781003308072-23

If a small, positive charge, labeled q in Figure 22.1, is located close to the right-hand side and near the middle of the region, it will have a force F on it due to the electric field. The magnitude of the force will be $F = Eq$. Remember that these electric field lines indicate the way a positive charge would be pushed in this region of space. Since the force is directly to the left and the displacement caused by this force is to the left, the electric field will do work on the charge that is $W = (Fd)$. Because work will be done on the charge, q, by the electric field, E, for this charge in this field, there is potential energy, U, with a value of $U = (Fd) = (Eqd)$. Since the magnitude of the test charge, q, is arbitrarily, the condition of the electric field set-up by the charge distributions, it make sense to measure its effect on a per-charge basis. Therefore, the electric potential is defined as the potential energy per charge in a region of space, and given in equation (22.1) as

$$V = \frac{U}{q} \hspace{5cm} (22.1)$$

The unit of electric potential is the Volt, which is just a Joule per Coulomb, $\frac{J}{C} = V$. This quantity of electric potential is commonly called "volt-age" in the same way people refer to the miles per gallon they get in their car as the "mileage". So, from this point forward electric potential will be referred to by its common name of voltage.

Like mechanical potential energy, gravitational potential energy, the point of zero potential energy, is arbitrary. For example, in Figure 22.2 the zero height is chosen in the middle of the ramp, to match the location of the zero voltage for the charge in Figure 22.1.

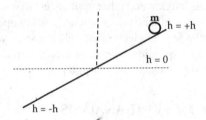

FIGURE 22.2   Comparison of voltage to gravitational potential energy.

For the charge, $+q$, in Figure 22.1 and the ball with a mass, m, in Figure 22.2, the electric field and the gravitational field, respectively, keep acting on the object even as it passes through the zero point. That is, the charge and the mass will keep moving through the location of zero potential energy since it was chosen at that point for convenience. The most common point of zero voltage, potential energy per charge, is the voltage of the earth's surface, which is called *ground* and is denoted with the symbol shown in Figure 22.3.

FIGURE 22.3   Ground symbol.

In many homes, this base, ground voltage, is literally an iron bar driven into the ground just outside your house and/or a cold-water pipe from the street, which is buried in the ground. It is the difference in voltage in Figure 22.1 from $+V$ to $-V$, like the difference in height in Figure 22.2 from $+h$ to $-h$, which are the quantities of physical significance. The difference in voltage measured from one point to another is equal to the work that would have to be done, per unit charge, against an electric field to move the charge from one point to another.

22.3.2 BATTERIES

The voltage of a battery is the result of work done in separating charge. A 9 V battery does more work per charge than a 1.5 V one but not necessarily more total work. Remember that voltage is potential energy per charge. This work done in a common battery is chemical in nature and results in the separation of charge from the positive side and the negative side of the battery.

The symbol for a battery is given in Figure 22.4.

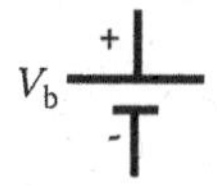

$$V_b \ \overset{+}{\underset{-}{\vdash}}$$

FIGURE 22.4   The circuit symbol for a battery.

This symbol will be used for all power supplies in this text. The two collinear line segments represent the terminals of the battery, with the longer of the two parallel bars the positive terminal. This is normally labeled with a + and the other with a –, as shown in Figure 22.4.

The voltage supplies used in labs and in most electronic devices achieve a voltage difference by the separation of charge, but not from a chemical reaction. The energy associated with the electrical outlet is generated in a power plant through a process called electrometric induction. This process involves the interaction of electric and magnetic fields moving charge around. The voltage supply in the lab simply reorganizes the energy in a way that constantly separates charge into positive and negative sides.

22.3.2.1   Electric Current (I)

Introduced in Chapter 6 in the context of magnetic force, the electric current is the rate at which electric charge flows past a given point in an electric circuit. It is expressed, as shown in equation (22.2), as the rate of change of charge,

$$I = \frac{\Delta q}{\Delta t}. \tag{22.2}$$

The units for current are coulombs per second, which is the Ampere, abbreviated as the "Amp" and given the symbol, A, so that $1 \ A = 1 \ \frac{C}{s}$

The moment that a voltage is put across a circuit, an electric field is created around the circuit at approximately the speed of light, but the charge in the circuit doesn't go that fast. In fact, the current through a circuit isn't very fast at all, because the electrons take a biased (forced) random walk, through the wires, colliding with nuclei and other electrons as they make their way around the circuit. The electrical current takes this random walk nature into account, because the current is a net result, not a theoretical value.

22.3.2.2   Resistance (R)

The resistance of a circuit element is a measure of how poorly an object conducts electricity. *Conducts* in the term that is used in electrical circuit analysis that allows electrons to move through the circuit. So, a circuit element, a piece of the circuit, that allows electrons to move through the circuit easily has a low resistance, and a circuit element that does not allow electrons to move through it easily has a high resistance. If you think about the word, it makes perfect sense. The bigger the value of resistance, the more poorly the circuit element allows charge to flow through itself. The symbol for resistance is an R, and the unit of resistance is the ohm (Ω). Small ceramic cylinders, with a coating of carbon and a wire sticking out each end, are produced for circuits to have specific values of resistance. A sketch of one of these resistors is given in Figure 22.5, in which the stripes on the resistor are color coded to give the manufactured resistance, with a first digit (1st), a second digit (2nd), a multiplier (M), and the tolerance (T) of the manufacturing level of certainty.

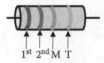

1st 2nd M T

FIGURE 22.5   A resistor.

These stripes indicate the value of the resistance and the tolerance of the known values, as given in Table 22.1.

TABLE 22.1
Resistor Color Code

Color	Black	Brown	Red	Orange	Yellow	Green	Blue	Violet	Gray	White
R =	0	1	2	3	4	5	6	7	8	9
Color			Red		Gold		Silver			None
Tolerance =			2%		5%		10%			20%

For example, a resistor with stripes from left to right of brown, black, red, and silver, is a resistor with a resistance of $10 \times 10^2 = 1000\ \Omega \pm 10\%$. The circuit element symbol for a resistor is given in Figure 22.6.

FIGURE 22.6   The circuit symbol for a resistor.

It looks like kinks in a water hose, which would cause a resistance in the flow of water through a hose. This symbol is also used in circuits for any object like a light bulb of a toaster element, if the resistance of the object is important to the analysis of the system, like in a home.

22.3.2.3   Ohm's Law

Given the circuit shown in the schematic diagram in Figure 22.7, if a battery with a voltage V is connected across a resistor, with a resistance, R, a current, I, will flow through the circuit.

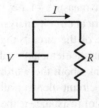

FIGURE 22.7   A simple circuit.

This is the language that is used in electrical circuit analysis, the voltage is measured across a resistor from one side to the other of the resistor, and the current flows through the wires and the resistors. Try to be aware of the preposition used to describe voltage and current. It is important in the description.

The relationship between voltage, current, and resistance was discovered by the German scientist Georg Simon Ohm in the early 1800s; thus, Ohm's Law is given in equation (22.3) form as,

$$V = IR. \tag{22.3}$$

The reason for this simple linear relationship is that in most electric circuits, as the voltage across (V) the circuit is increased the current (I) flowing through the circuit increases by the same proportion. Therefore, if the voltage across the resistor in Figure 22.7 is plotted vs the current flowing through the circuit, a graph similar to the one shown in Figure 22.8 is generated.

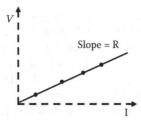

FIGURE 22.8 Graph of voltage across a circuit element as a function of the current through the same element.

The slope of the line through the data points is the resistance (R) of the resistor. An increase of the voltage across a circuit, causes more charge to move from a higher to a lower voltage. As mentioned above, the unit of resistance is the ohm (Ω), which is the ratio of a volt (V) to an amp (A), $\Omega = \frac{V}{A}$. If the material of which the resistor is made obeys Ohm's Law, then the resistance R is a constant, meaning that its value is the same for different voltages and currents. These materials are called ohmic.

Example 1

A current of 2.0 A flows through a resistor that is connected across a 6.0 V battery. What is the resistance of the resistor?

Solution: Generate a schematic diagram, like the one in Figure 22.9.

FIGURE 22.9 Example 1 schematic diagram.

Write out Ohm's Law: $V = IR$
Solve and plug in the values: $R = \frac{V}{I} = \frac{6\ V}{2\ A} = 3\ \Omega$

22.3.2.4 Electrical Power

In Chapter 16, power was defined as the rate at which energy is transformed from one form of energy into another form of energy, in equation (16.10) as $P_{avg,01} = \frac{\Delta E_{01}}{\Delta t_{01}}$, and in equation (22.1) voltage is defined as $V = \frac{U}{q}$ and in equation (22.2) current is defined as $I = \frac{\Delta q}{\Delta t}$, then power can be expressed as the product of current through and voltage across a circuit element in equation (22.4) as,

$$P = IV. \tag{22.4}$$

In addition, by inserting IR for V from Ohm's Law into equation (22.4), an expression of power in terms of current and resistance can be given in equation (22.5) as,

$$P = I^2R. \tag{22.5}$$

The unit of power is the watt, $W = \frac{J}{s}$.

Although equations (22.4) and (22.5) are interchangeable, it is customary to use equation (22.4) to compute the power supplied to a circuit by a battery or another source of voltage. On the other hand, the power dissipated, or used, by a resistor is usually computed using equation (22.5).

Another antiquated name for a supplied voltage is the electromotive force, or emf, which is assigned the symbol ($\mathcal{E}$). This emf is still used in some physics texts today to indicate the theoretical voltage of a battery or other power sources, but it is still just electric potential or voltage. It came from the idea that there must be an electrostatic force that pushes positive charges through the circuit.

22.3.2.5 Electricity – Water Analogy

The flow of electricity through the wires in a circuit is similar to the flow of water through the pipes in a system, like the one in Figure 22.10.

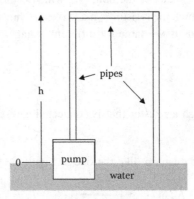

FIGURE 22.10 Water analogy of electricity.

For water in a pipe, a pump provides a pressure difference that lifts the water to a height, h, above the zero point (0) at the surface of the water. Each gram of water has a potential energy, $U = (mgh)$, so the potential energy per mass is $U/m = gh$, which is analogous to the measure of voltage. The higher the water is lifted above the surface, the faster it will flow through the pipes down to the surface of the water. The flow of water in grams/s is analogous to the electrical current in C/s. It is obvious from this analogy that the size of the pipes will also have an effect on the flow rate of water through the system. Given the same potential energy per mass of the water, more water will flow through larger diameter pipes than pipes at the same height with smaller diameter. The smaller pipes create a greater resistance to the flow of water through the pipes, in the same way the size of wires affects the electrical resistance of the wires. This analogy is common in the study of electrical circuits and will be carried throughout this chapter.

22.4   KIRCHHOFF'S LAWS

22.4.1   KIRCHHOFF'S CURRENT LAW (KCL)

Conservation of electric charge states that the net charge flowing into any part of an electrical circuit is equal to the net charge flowing out of that same part. Since the flow of electrical charge is electrical current (I), conservation of charge can be written in equation (22.6a) as,

$$\sum I_{in} = \sum I_{out}.$$ (22.6a)

This is Kirchhoff's Current Law. Known to physicists and electrical engineers as the "Junction Law" or "KVL". It is also common to write equation (22.6b) as

$$\sum I_{in} - \sum I_{out} = 0.$$ (22.6b)

At a point in a circuit where charge has several possible paths to travel, KCL gives an expression for the flow of charge. For example, in Figure 22.11 a current, I, flows through a wire from left to right. At the point n, it comes to a junction, n.

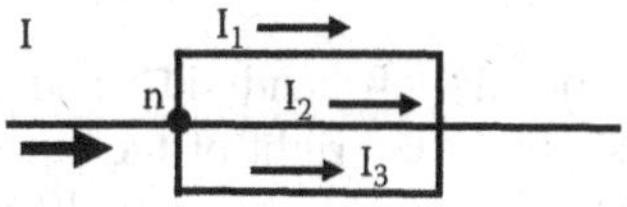

FIGURE 22.11   Kirchhoff's Current Law (KCL).

In electrical circuit analysis, these points are known as nodes, thus the letter n. At this node, the current splits up into three paths labeled I_1, I_2, and I_3. The total charge flowing into a junction must be the same as the total charge flowing out of the junction. Thus, $I = I_1 + I_2 + I_3$ or $I - I_1 - I_2 - I_3 = 0$.

22.4.2   KIRCHHOFF'S VOLTAGE LAW

Kirchhoff's Voltage Law (KVL) is conservation of energy applied to an electrical circuit. Since voltage is the energy per charge at a point in an electrical circuit, KVL states that the sum of the voltage supplied around a closed loop must be equal to the sum of the voltage dissipated around the loop, so that the net voltage around the loop is zero. That is, the voltage in (V_{in}) is equal to the voltage out (V_{out}). This is expressed in equation (22.7a) as,

$$\sum V_{in} = \sum V_{out}.$$ (22.7a)

This is Kirchhoff's. Known to physicists and electrical engineers as "KVL". It is also common to write equation (22.7b) as

$$\sum V_{in} - \sum V_{out} = 0$$ (22.7b)

In practice the V_{out} is commonly replaced with the expression that is specific to the circuit element that is dissipating the voltage. For a resistor, it is Ohm's Law, so the right-hand side of equation (22.6) is commonly a sum of products of current and resistance.

In Figure 22.12, the three currents, I_1, I_2, and I_3, are drawn in the circuit in the direction it is assumed the current will flow. It is not important which way the currents are drawn, only that once they are drawn in the circuit diagram they must remain in the same direction for the entire analysis. If by chance the current is labeled in a direction opposite the way the current is flowing, in that section of the circuit the numerical answer will come out as negative. So, choose a direction of current flow and be consistent throughout the problem.

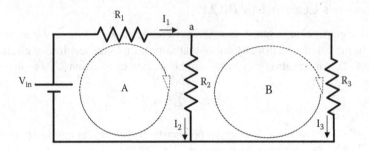

FIGURE 22.12 Electrical circuit for KVL.

For each of the three loops, A, B, and C, the voltages around each loop sum to zero. For loop A, the voltage supplied is V_{in} and the two voltage drops are across R_1 and R_2, so KVL for this loop is:

$$V_{in} = R_1 I_1 + R_2 I_2$$

Notice that the voltage supplied is on the left-hand side, and the voltage dissipated, known as "voltage drops" across the resistors, are on the right of the equal sign.

For loop B, there is no voltage supplied, and only two voltage drops are across R_1 and R_2. So, KVL for this loop is:

$$0 = R_2(-I_2) + R_3 I_3$$

Notice that there is a negative sign in front of I_2 in the first term on the right-hand side of the previous expression. The reason for this negative sign is the clockwise direction of loop B, and the current I_2 was chosen to flown downward through $R2$, so the loop and the current are in opposite directions.

In addition, the node a provides another equation of further information through KCL.

$$I_1 = I_2 + I_3 \quad \text{or} \quad I_1 - I_2 - I_3 = 0$$

22.4.3 CIRCUIT ANALYSIS EXAMPLE

Example 1

Find the current through each of the resistors in the circuit in Figure 22.13.

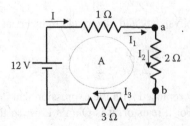

FIGURE 22.13 Electrical circuit for example 1.

Solution

Step 1. Label the currents and the loops in the diagram, as shown in Figure 22.13.

Step 2. Write out the KVL and KCL equations for the circuit.

    A. KVL for loop A is: $V_{in} = R_1I_1 + R_2I_2 + R_3I_3$

    B. KCL for the two nodes a and b are: $I_1 = I_2$ and $I_2 = I_3$, respectively.

        Therefore: $I_1 = I_2 = I_3$, so all the I's are the same; thus, they can all be considered I. So, KVL for loop A is: $V_{in} = R_1I + R_2I + R_3I$.

Step 3. Plug in the numbers and solve for the unknowns. $V_{in} = R_1I + R_2I + R_3I$

$$12 \ V = (1 \ \Omega)I + (2 \ \Omega)I + (3 \ \Omega)I = (6 \ \Omega)I$$

Solving for I gives: $I = (12 \ V)/(6 \ \Omega) = 2 \ A$

Plugging back into Ohm's Law for each resistor gives:

$$
\begin{aligned}
V_1 &= I_1R_1 = (1 \ \Omega) \ (2 \ A) = 2 \ V \\
V_2 &= I_2R_2 = (2 \ \Omega) \ (2 \ A) = 4 \ V \\
V_3 &= I_3R_3 = (3 \ \Omega) \ (2 \ A) = 6 \ V
\end{aligned}
$$

$V_1 + V_2 + V_3 = 12 \ V$ *(This is the same as the voltage supplied, so KVL is satisfied.)*

Example 2

Find the current through each of the resistors in the circuit in Figure 22.14.

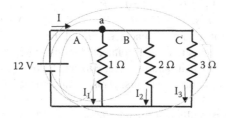

FIGURE 22.14 Electrical circuit for example 2.

Solution

Step 1. Label the currents and the loops in the diagram, as shown in Figure 22.14.

Step 2. Write out the KVL and KCL equations for the circuit.

    A. KVL for loop A is: $V_{in} = R_1I_1$

    B. KVL for loop B is: $V_{in} = R_2I_2$

    C. KVL for loop C is: $V_{in} = R_3I_3$

    D. KCL for node a is: $I = I_1 + I_2 + I_3$.

        It is clear for resistors connected with all the positive sides connected with a wire and all the negative sides connected with a wire, that every resistor has the same voltage across it and the total current, I, coming out of the battery is the sum of the individual currents. This arrangement of resistors is called *in-parallel*.

Step 3. Plug in the numbers and solve for the unknowns.

$$12 \ V = (1 \ \Omega)I_1, \quad 12 \ V = (2 \ \Omega)I_2, \quad and \quad 12 \ V = (3 \ \Omega)I_3$$

Solving for each I gives: $I_1 = 12\ A$, $I_2 = 6\ A$, and $I_3 = 4\ A$

Find the total current, I, with the KCL:

$$I = I_1 + I_2 + I_3 = 12\ A + 6\ A + 4\ A = 22\ A$$

Example 3

Find the current through each of the resistors in the circuit in Figure 22.15, which is the circuit in Figure 22.12 with values for the resistances and the battery voltage.

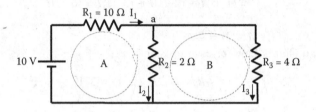

FIGURE 22.15   Electrical circuit for KVL.

Solution

Step 1. Label the currents and the loops in the diagram, as shown in Figure 22.15.
Step 2. Write out the KVL and KCL equations for the circuit.
  A. KVL for loop A is: $V_{in} = R_1 I_1 + R_2 I_2$
  B. KVL for loop B is: $0 = R_2(-I_2) + R_3 I_3$
  C. KCL for node a is: $I_1 - I_2 - I_3 = 0$.
        It is clear for resistors connected with all the positive sides connected with a wire and all the negative sides connected with a wire, that every resistor has the same voltage across it, and the total current, I, coming out of the battery is the sum of the individual currents. This arrangement of resistors is called *in-parallel*.
Step 3. Plug in the numbers and solve for the unknowns.
  A. KVL for loop A is: $10\ V = (10\ \Omega)\ I_1 + (2\ \Omega)\ I_2$
  B. KVL for loop B is: $0 = (2\ \Omega)\ (-I_2) + (4\ \Omega)\ I_3$
  C. KCL for node a is: $I_1 - I_2 - I_3 = 0$.

The plan is to solve both equations A and B in terms of I_2 and then plug them both into equation C to find I_2.

Solving equation A for I_1 gives: $I_1 = [(10\ V)/(10\ \Omega)] - [(2\ \Omega)/(10\ \Omega)]$ so $I_1 = 1\ A - .2\ I_2$
Solving equation B for I_3 gives: $I_3 = [(2\ \Omega)/(4\ \Omega)]I_2$ so $I_3 = (0.5)I_2$
Plugging in the expressions for I_1 and I_3 found in the previous two steps into equation C: $I_1 - I_2 - I_3 = 0$, gives: $(1\ A - .2\ I_2) - I_2 - (0.5)I_2 = 0$. This can be solved for I_2 as: $1\ A = .2\ I_2 + I_2 + 0.5 I_2$ so $1\ A = 1.7\ I_2$. Therefore, $1\ A = 1.7\ I_2$ so $I_2 = 0.588\ A$.
Plugging back into $I_1 = 1\ A - .2\ I_2 = 1A - .2(0.588\ A) = 0.882\ A$.
Plugging back into $I_3 = (0.5)I_2 = (0.5)\ (0.588\ A) = 0.294\ A$.
The final answers for the currents are:

$$I_1 = 0.882\ A, \quad I_2 = 0.588\ A, \quad \text{and } I_3 = 0.294\ A$$

As a check, find the voltage across each of the resistors with Ohm's Law.

$$V_1 = R_1 I_1 = (10\ \Omega)\ (0.882\ A) = 8.82\ V$$
$$V_2 = R_2 I_2 = (2\ \Omega)\ (0.588\ A) = 1.18\ V$$
$$V_3 = R_3 I_3 = (4\ \Omega)\ (0.294\ A) = 1.18\ V$$

Notice that loop A adds up to ($8.82\ V + 1.18\ V = 10\ V$) supplied and that loop B is equal to each other at $1.18\ V$ each.

22.5   COMBINING RESISTORS

The analysis of a circuit involves the determination of the voltage across, and the current through, circuit elements in that circuit. All circuit analysis can be done with Kirchhoff's Laws, but in some cases the algebra of solving multiple equations with multiple unknowns can be difficult. There is another method that involves replacing a combination of resistors with a single equivalent resistor that will not change the voltage across or the current through any circuit elements in the circuit. The process results in a circuit that is easier to analyze, and the results of its analysis apply to the original circuit.

22.5.1   RESISTORS IN SERIES

Resistors in series are connected so that the side of one resistor that is closest to the positive side of the power supply is connected to the side of the resistor that is closest to the negative side of the same power supply. Thus, the resistors are connected so that current flows out of one of them and into the next one, as shown in Figure 22.16.

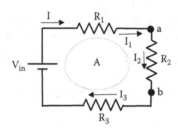

FIGURE 22.16   Electrical circuit of resistors in series.

By KCL, the current through the entire circuit is the same for each of the resistors, so

$$I = I_1 = I_2 = I_3$$

By KVL for loop A, the voltage supplied by the power supply must equal the sum of the voltage drops across the resistors:

$$V_{in} = V_1 + V_2 + V_3.$$

By Ohm's Law ($V = IR$), where the equivalent of the entire circuit's resistance is labeled R_{eq}:

$$IR_{eq} = I_1 R_1 + I_2 R_2 + I_3 R_3$$

Since all the currents are the same throughout the circuit, the equivalent resistance of the circuit of series resistors is:

$$R_{eq} = R_1 + R_2 + R_3$$

Thus, the equivalent resistance of resistors connected in series is the sum of their resistances, as shown in equation (22.8) as:

$$R_{series} = \sum_i R_i \tag{22.8}$$

22.5.2 RESISTORS IN PARALLEL

Resistors in parallel are connected so that the positive sides of all the resistors are connected together and all the negative sides of the resistors are connected together, as shown in Figure 22.17.

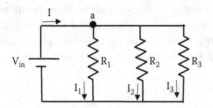

FIGURE 22.17   Electrical circuit of resistors in parallel.

Thus, the current flowing through the circuit runs in parallel. By KCL at node *a*, at the top of the circuit, the main current splits up into the current in each of the branches of the circuit, so that

$$I = I_1 + I_2 + I_3$$

By KVL for each of the three loops which include V_{in} and one of the resistors, the voltage supplied by the power supply must equal the voltage drops across each of the resistors:

$$V_{in} = V_1 = V_2 = V_3.$$

Starting with KCL and using Ohm's Law ($I = V/R$), where the equivalent of the entire circuit's resistance is labeled R_{eq}:

$$\frac{V_{in}}{R_P} = \frac{V_1}{R_1} + \frac{V_2}{R_2} + \frac{V_3}{R_3} + \dots$$

Since all the voltages are the same across each element in the circuit, the equivalent resistance of the circuit of parallel resistors is found with

$$\frac{1}{R_P} = \frac{1}{R_1} + \frac{1}{R_2} + \frac{1}{R_3} + \dots$$

Which can be summarized as equation (22.9) as:

$$\frac{1}{R_P} = \sum_i \frac{1}{R_i}. \tag{22.9}$$

Example 4

Find the voltage across, and the current through, each of the resistors in Figure 22.18, given that $V_{in} = 12$ V, $R_1 = 2\ \Omega$, $R_2 = 4\ \Omega$, $R_3 = 5\ \Omega$, and $R_4 = 3\ \Omega$.

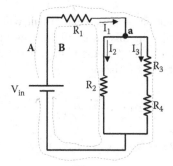

FIGURE 22.18   Electrical circuit of example 4.

Solution 1

The first solution employs Kirchhoff's Laws.

  Step 1. Sketch the loops A and B, the node a, the currents (I_1, I_2, and I_3).
  Step 2. Write out the KVL and KCL equations for the circuit and use the numbers directly.
    A.   KVL for loop **A** is: $12\ V = (2\ \Omega)(I_1) + (5\ \Omega)I_3 + (3\ \Omega)I_3$
    B.   KVL for loop **B** is: $12\ V = (2\ \Omega)(I_1) + (4\ \Omega)I_2$
    C.   KCL for node a is: $I_1 - I_2 - I_3 = 0$.
  Step 3. Solve for the unknowns using a technique similar to the one used in example 3 or using advanced techniques like Row Reduced Echelon Form (RREF).

The final answers for the currents are: $I_1 = 2.57\ A$, $I_2 = 1.71\ A$, and $I_3 = 0.86\ A$

Solution 2

The second solution employs the concepts of equivalent resistance.

  Step 1. Find the equivalent resistance of the circuit. Looking at the circuit in Figure 22.18, the equivalent resistance must be found in parts because of the two different ways the equivalent resistance of resistors in series and parallel are computed.
    First, find $R_{34} = R_3 + R_4 = 3\ \Omega + 5\ \Omega = 8\ \Omega$, because the R_3 and R_4 are in series with each other and only each other.
    Second, find the resistance of the parallel combination of R_2 and R_{34}:

$$\frac{1}{R_{234}} = \frac{1}{R_2} + \frac{1}{R_{34}} = \frac{1}{4\Omega} + \frac{1}{8\Omega} = \frac{2}{8\ \Omega} + \frac{1}{8\Omega} = \frac{3}{8\ \Omega}\ \text{so,}\ \ R_{234} = \frac{8}{3}\Omega$$

Lastly, add R_1 to R_{234} to get the equivalent resistance, R_{eq}, since R_1 is in series with the parallel combination of R_{234}.

$$R_{eq} = R_1 + R_{234} = 2\Omega + \frac{8}{3}\Omega = \frac{6}{3}\Omega + \frac{8}{3}\Omega = \frac{14}{3}\Omega = 4\frac{2}{3}\Omega$$

  Step 2. Use Ohm's Law to find the current through the entire circuit, which in this circuit flows through R_1.

$$I_1 = \frac{V_{in}}{R_{eq}} = \frac{12V}{4\frac{2}{3}\Omega} = 2.57\ A$$

Notice that this value agrees with the I_1 found in the first solution of this circuit.

Step 3. At this point in the solution process, each circuit will pose a different set of steps depending on the circuit. In this case, all the current flows through the parallel section with a resistance of R_{234}, so the voltage across R_{234} is found by multiplying the entire current, I_1, by this resistance.

$$V_{234} = I_1 R_{234} = (2.57 \ A)\left(\frac{8}{3}\Omega\right) = 6.86V$$

Since this entire voltage drops across R_2, I_2 can be found as:

$$I_2 = \frac{V_{234}}{R_2} = \frac{6.86}{4 \ \Omega} = 1.71 \ A$$

Notice that this value agrees with the I_2 found in the first solution of this circuit.

This last one is little more difficult. Notice that that the entire parallel voltage drops across R_{24}, so I_3 can be found as:

$$I_2 = \frac{V_{234}}{R_{34}} = \frac{6.86}{8 \ \Omega} = 0.86 \ A$$

Notice that this value agrees with the I_3 found in the first solution of this circuit.

Using either technique, the last step is employing Ohm's Law to find the voltage across each of the resistors.

$$R_1 \ = \ 2 \ \Omega, \quad R_2 = 4 \ \Omega, \quad R_3 = 5 \ \Omega, \quad \text{and } R_4 = 3 \ \Omega.$$
$$I_1 \ = \ 2.57 \ A, \quad I_2 = 1.71 \ A, \quad \text{and } I_3 = 0.86 \ A$$

$$
\begin{aligned}
V_1 \ &= \ R_1 I_1 = (2 \ \Omega) \ (2.57 \ A) = 5.14 \ V \\
V_2 \ &= \ R_2 I_2 = (4 \ \Omega) \ (1.71 \ A) = 6.84 \ V \\
V_3 \ &= \ R_3 I_3 = (5 \ \Omega) \ (0.86 \ A) = 4.30 \ V \\
V_4 \ &= \ R_3 I_3 = (3 \ \Omega) \ (0.86 \ A) = 2.58 \ V
\end{aligned}
$$

Notice that voltages around both loop A and B add up to approximately 12 V.

(5.14 V + 6.84 V = 11.98 V = 12 V) and *(5.14 V + 4.30 V + 2.58 V = 12.02 V = 12 V)*. This is a good check of the solution.

22.5.3 RESISTIVITY

If an object is made out of a single type of material, its resistance depends on the shape of the object and the type of material. The quality of the material that affects the resistance is called the *resistivity of the material.*

A cylindrical object made of a single material, such as a manufactured resistor, is depicted in Figure 22.19.

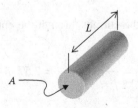

FIGURE 22.19   A cylindrical object.

Most of the materials from which resistors are made are ohmic, which means that they obey Ohm's Law over a large range of temperatures. For these materials, the resistance is inversely proportional to the cross-sectional area, A, directly proportional to the length, L of the object; therefore, the resistance, R, of the object can be represented as equation (22.10) as:

$$R = \rho \frac{L}{A}.$$
(22.10)

where ρ is the resistivity of the substance of which the resistor is made.

The values of resistivity for several common materials are provided in Table 22.2:

TABLE 22.2
Resistivity of Some Materials

Material	Resistivity ρ
Silver	1.6×10^{-8} Ω·m
Copper	1.7×10^{-8} Ω·m
Gold	2.4×10^{-8} Ω·m
Aluminum	3×10^{-8} Ω·m
Tungsten	5.6×10^{-8} Ω·m
Seawater	0.25 Ω·m
Axoplasm	1.1 Ω·m
Cell Membrane	1×10^{7} Ω·m
Myelin	1×10^{7} Ω·m
Rubber	1×10^{13} Ω·m
Glass	1×10^{10} to 1×10^{14} Ω·m

Note that in SI units, the units of resistivity are ohms *times* meters.

22.6   ANSWER TO CHAPTER QUESTION

Your home is set up as a parallel circuit so that as more electronic devices are plugged in, the equivalent resistance of your home decreases, which allows more current to flow into your home. The limit is set by the service to your house and the breakers, which limit the current to safe values.

22.7   QUESTIONS AND PROBLEMS

22.7.1   MULTIPLE CHOICE QUESTIONS

1. Given that all the lamps in a house have the same voltage across the light bulbs, which light bulb, a 40 W bulb or a 100 W bulb, has a higher resistance?
   A.   40 W bulb
   B.   100 W bulb
   C.   Both have the same resistance.

2. Given a circuit made up of a battery and two resistors, R_1 and R_2, all of which are in series with each other. If the resistance of resistor R_2 is twice that of R_1, is the current through R_2 greater than, less than, or equal to the current through R_1?
   A.   greater than
   B.   less than
   C.   equal to

3. A circuit is produced that consists of a battery and two resistors, R_1 and R_2, all in series with each other. If the resistance of resistor R_2 is twice that of R_1, is the voltage across R_2 greater than, less than, or equal to the voltage across R_1?
   A. greater than
   B. less than
   C. equal to

4. A circuit is produced that consists of a battery and two resistors, R_1 and R_2, all in parallel with each other. If the resistance of resistor R_2 is twice that of R_1, is the current through R_2 greater than, less than, or equal to the current through R_1?
   A. greater than
   B. less than
   C. equal to

5. Consider a circuit consisting of a battery and two resistors, R_1 and R_2, all of which are in parallel with each other. If the resistance of resistor R_2 is twice that of R_1, is the voltage across R_2 greater than, less than, or equal to the voltage across R_1?
   A. greater than
   B. less than
   C. equal to

Questions 6–10 refer to the circuit in Figure 22.20.

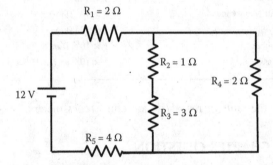

FIGURE 22.20 Multiple choice questions 6–10 and problems 9 & 10.

6. Is the current through R_1 greater than, less than, or equal to the current through R_5?
   A. greater than
   B. less than
   C. equal to

7. Is the current through R_2 is greater than, less than, or equal to the current through R_4?
   A. greater than
   B. less than
   C. equal to

8. Is the voltage across R_3 greater than, less than, or equal to the voltage across R_4?
   A. greater than
   B. less than
   C. equal to

9. Is the voltage across R_2 greater than, less than, or equal to the voltage across R_3?
   A. greater than
   B. less than
   C. equal to

10. Which is the resistor with the greatest voltage drop across it in this circuit?
    A. R_1
    B. R_2
    C. R_3
    D. R_4
    E. R_5
    F. All voltages are the same.

22.7.2 PROBLEMS

1. For the circuit in Figure 22.21, the voltage of the battery is 4 V, and the resistance of the resistor is 2 Ω. Find the:
   a. voltage across the resistor,
   b. current through the resistor, and
   c. power dissipated by the resistor.

FIGURE 22.21   Problem 1.

2. For the circuit in Figure 22.22, the voltage of the battery is 4 V, and the resistance of each of the two resistors is 2 Ω. Find the:
   a. voltage across each resistor,
   b. current through each resistor, and
   c. power dissipated by each resistor.

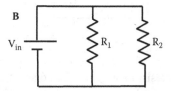

FIGURE 22.22   Problem 2.

3. For the circuit in Figure 22.23, the voltage of the battery is 12 V, the resistance of R_1 is 2 Ω, and the resistance of R_2 is 4 Ω. Find the:
   a. voltage across each resistor,
   b. current through each resistor, and
   c. power dissipated by each resistor.

FIGURE 22.23   Problem 3.

4. The resistances of the three resistors in Figure 22.24 are $R_1 = 2$ Ω, $R_2 = 4$ Ω, and $R_3 = 8$ Ω. The battery supplies a constant voltage of 12 V. What is the total current flowing out the battery?

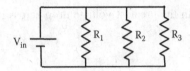

FIGURE 22.24   Problem 4.

5. The resistances of the three resistors in Figure 22.25 are $R_1 = 2\ \Omega$, $R_2 = 12\ \Omega$, and $R_3 = 4\ \Omega$. The battery supplies a constant voltage of $V_{in} = 10$ V. What is the current through and voltage across R_3?

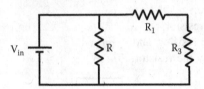

FIGURE 22.25   Problem 5.

6. The resistances of the four resistors in Figure 22.26 are $R_1 = 20\ \Omega$, $R_2 = 10\ \Omega$, $R_3 = 15\ \Omega$, and $R_4 = 25\ \Omega$. The battery supplies a constant voltage of $V_{in} = 10$ V. What is the current flowing out of the battery?

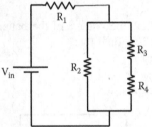

FIGURE 22.26   Problems 6 and 7.

7. The resistances of the four resistors in Figure 22.26 are $R_1 = 20\ \Omega$, $R_2 = 10\ \Omega$, $R_3 = 15\ \Omega$, and $R_4 = 25\ \Omega$. The battery supplies a constant voltage of $V_{in} = 10$ V. What is the current through and voltage across R_3?

8. The resistances of the four resistors in Figure 22.27 are $R_1 = 100\ \Omega$, $R_2 = 500\ \Omega$, $R_3 = 300\ \Omega$, and $R_4 = 700\ \Omega$. The battery supplies a constant voltage of $V_{in} = 10$ V. What is the current through and voltage across R_4?

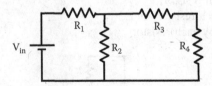

FIGURE 22.27   Problem 8.

9. For the circuit in Figure 22.20, find the current flowing through R_1.
10. For the circuit in Figure 22.20, find the current flowing through R_3.

23 Capacitance

23.1 INTRODUCTION

In the previous chapter, the relationship between the current flowing through a circuit element and the voltage across the element was defined as the resistance of the circuit element. The rules and methods used to study these devices are based on conservation of energy and conservation of charge. If the circuit element is designed to collect charge instead of letting it flow through the circuit, its ability to store charge is appropriately defined as its capacitance. It turns out that these devices that collect charge do so to store energy in the circuit; therefore, the analysis of these devices and their role in a circuit are centered around their energy storage and energy dissipation capabilities. Thus, the focus of this chapter is on the capacitance of circuit elements and the methods applied to study the roles of these devices in an electrical circuit.

23.2 CHAPTER QUESTION

How can a battery, with a rating of only a few volts, supply the large burst of energy delivered by a portable defibrillator? The answer to this question lies in the concept of electrical capacitance and will be answered at the end of the chapter.

23.3 CAPACITANCE

In the previous chapter, a source of voltage was connected across a resistor, and a current flowed through the resistor. The amount of current was proportional to the voltage and inversely proportional to the resistance. In this chapter, the device at the center of the discussion does not allow a proportional current flow when a voltage is connected across it. Instead, it stores energy in the form of displaced charge.

One of the simplest examples of this type of object is a parallel plate capacitor. Depicted in Figure 23.1, this device is simply two conducting plates, each with an area A and usually made of metal, positioned so that they are parallel to each other and separated from one another, by a distance d, so that charge cannot cross from one plate to another.

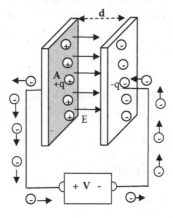

FIGURE 23.1 Parallel plate capacitor.

In the simplest case, the plates are separated by air, which does not conduct electricity well. When this device is connected across a voltage supply, like a battery, negative charge will build up

DOI: 10.1201/9781003308072-24

on the plate connected to the negative side of the supply, and positive charge will build up on the other plate. The voltage supply does work pulling electrons from the positive plate and pushing electrons onto the negative plate, as depicted in Figure 23.1. The electric field that develops between the plates, as shown in Figure 23.1, is created by the charge distribution, and it provides a way to analyze the capacitor in terms of its dimensions.

With a power supply, like a battery, connected across a capacitor, such as the box labeled with a, + V −, in Figure 23.1, charge is displaced in an unbalanced arrangement, with more positive charge on one plate and more negative charge on the other plate. It is not the same exact particles that are moved from one plate to another, but it is the same amount of charge, q. This difference in electric potential is the voltage across the capacitor, V. The ratio of the amount of displaced charge moved off one plate and onto the other plate and the voltage across the capacitor is the capacitance, C, of the device, as defined in equation (23.1) as

$$C = \frac{q}{V}.$$

(23.1)

The unit for capacitance is the Farad, which is denoted as F and equals a Coulomb per Volt. Like the resistance and the unit of an Ohm (Ω) for a resistor, the capacitance of a device is set by the physical properties of the device, and the unit of the Farad has meaning only as the ratio of charge and voltage. Thus, the capacitance is a property of the device, and the charge to voltage (q/V) ratio is satisfied for that device. If a specific voltage, V, is connected across a device with a given capacitance, C, a quantity of charge, q, will be displaced onto the plates of the capacitor. A helpful analogy of the relationship between the area of the plates of a capacitor and the capacitance, is a bucket of water. As shown in Figure 23.2, if two buckets with two different areas, A_1 and A_2, have their bottoms filled to the same height h of water, the bucket with the larger area, A_1, will hold a larger volume of water; that is, it will have a larger capacity.

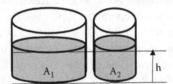

FIGURE 23.2   Two buckets with different areas and the same height of water in them.

In this analogy, the height of the water level is analogous to the voltage across the capacitor, the area of the bucket is related to the capacitance of the device, and the amount of water is related to the charge on the plates.

Example

Find the charge on the plates of a 5 μF capacitor with 1 kV across it. Rearrange equation (23.1) and note that a microfarad (mF) is 1×10^{-6} F and a kilovolt (kV) is 1×10^3 V

$$q = CV = (5 \times 10^{-6} \ F)(1 \times 10^3 \ V) = 0.005 \ C = 5 \ mC$$

The capacitance of parallel-plate capacitor, such as the one in Figure 23.1, is given by equation (23.2) as

$$C = \kappa \varepsilon_o \frac{A}{d}$$

(23.2)

Please see the Appendix of this chapter for the derivation of equation (23.2). In this expression, C is the capacitance, d is the separation distance between the plates, A is area of one face of one of the plates, ε_o is a constant called the permittivity of free space with a value of $\varepsilon_o = 8.85 \times 10^{-12} \frac{C^2}{N \cdot m^2}$, and κ is the dielectric constant related to the material's ability to support an electric field between the plates. The material between the plates also affects the ability of the device to separate charge, and this is the purpose of the dielectric constant, κ, in equation (23.2). With the effect on the electric field of a vacuum set at a value of $\kappa = 1$, the effect on an electric field due to the material between the plates is scaled relative to a vacuum. The dielectric constants of some materials used in manufactured capacitors and in our bodies are provided in Table 23.1. As evident from the values in this table, these materials enhance the electric field between the plates of the capacitor.

TABLE 23.1
Dielectric Constants of Some Materials

Substance	Dielectric Constant
Vacuum	1
Air	1
Waxed Paper	3
Titanium Dioxide	114
Aluminum Oxide	7
Myelin Sheath Membrane	7

Example

Compute the capacitance of a parallel-plate capacitor, with plates that have an area of 0.04 m^2 and are separated by 2 cm of air.
Use equation (23.2):

$$C = \kappa \varepsilon_o \frac{A}{d} = 1\left(8.85 \times 10^{-12} \frac{C^2}{N \cdot m^2}\right) \frac{0.04 \; m^2}{0.02 \; m} = 1.77 \times 10^{-11} F$$

This is the area of a square capacitor plate with 20 cm long slides, so the capacitance is commonly a small value. Therefore, capacitance is in units of picofarads (1 pF = 1×10^{-12} F) or nanofarads (1 nF = 1×10^{-9} F). Equation (23.2) is a good approximation as long as the plate separation d is small compared to the area of the plates. This is because the electric field is uniform over most of the interior area of the plates, but is not uniform near the edges of the plates, as depicted in Figure 23.3.

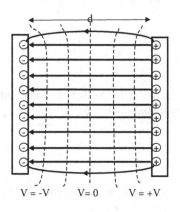

FIGURE 23.3 The electric field between the plates of a capacitor.

If the field in this figure looks familiar, it is the one used to explain the concept of voltage in the previous chapter. It is redrawn here in this context, because the electric field between the plates of the capacitor plays an important role in the explanation of capacitance.

From equation (23.2) it is clear that a capacitor with a larger area has a larger capacitance. If the distance between the plates is increased, the capacitance of the device decreases.

23.3.1 ENERGY STORED IN A CAPACITOR

Moving charge from one initially neutral capacitor plate to the other is called *charging the capacitor*. When a capacitor is charged, it is storing energy in the capacitor, because the voltage source, like a battery, moves some charge over a distance from one plate to another. Providing a conducting path for the charge to go back to the plate from which it came is called *discharging the capacitor*. If the capacitor is discharged through a device like an electric motor, work can be done by the energy stored in the capacitor.

Since voltage (V) is potential energy per charge, the product of voltage and charge has units of energy. Since the amount of charge separated across the plates of a capacitor increases linearly with an increase in the voltage across the capacitor, a greater voltage means more charge separated. A plot of the charge separated (q) on a plate as a function of the voltage (V) across the plates of the capacitor is shown in Figure 23.4.

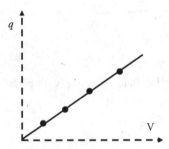

FIGURE 23.4 Graph q vs V for capacitor.

The area under the line formed by this graph is the potential energy stored in a capacitor, which is given in equation (23.3) as

$$U = \frac{1}{2}Vq \tag{23.3}$$

Since $C = q/V$ the potential energy can also be written in terms of V and C as shown in equation (23.4) as

$$U = \frac{1}{2}CV^2 \tag{23.4}$$

So, the C is like the spring constant, and the voltage is analogous to the displacement, s, of the spring from equilibrium in the potential energy expression of a spring given in equation (16.4) as $U_s = \frac{1}{2}k\,s^2$.

Example

Compute the energy stored in a capacitor with a capacitance of 8 μF and a voltage of 2 kV across the plates.

Rearrange equation (23.4):

$$U = \frac{1}{2}CV^2 = \frac{1}{2}(8 \times 10^{-6}F)(2 \times 10^3)^2 = 32 \; J$$

The energy is stored in the separation of the charge that forms the electric field, E, between the plates of the capacitor in Figure 23.1 or in Figure 23.3.

23.3.2 THE ELECTRIC FIELD OF A CAPACITOR

The electric field between the plates of charged parallel plates, in Figure 23.3, provides a way to understand how the capacitor stores energy. The electric field is the force per unit charge. Thus, multiplying the electric field strength (E) times the plate separation (d) gives the work per unit charge, which is by definition, the change in voltage, as

$$Ed = \left(\frac{F}{q}\right) d = \frac{(F \; d)}{q} = \frac{W}{q} = \frac{U}{W} = V$$

In summary, the expressions that relate the electric field to the energy and voltage is helpful in solving some problems, so they are summarized in equation (23.5) as

$$V = \frac{U}{q} = Ed \tag{23.5}$$

Example

Compute the electric field strength between the parallel plates of a capacitor that are separated by 2 cm when a voltage of 6 kV is put across the plates.

Remember that 6 kV = 6000 V and 2 cm is 0.02 m and rearrange equation (23.5):

$$E = \frac{V}{q} = \frac{6000 \; V}{0.02 \; m} = 300{,}000 \; \frac{V}{m} = 300{,}000 \; \frac{N}{C}$$

In most circuits, the capacitors are not actually two parallel plates but layers of conductors separated by layers of insulators, all rolled up in a small package. A simple way of visualizing this is shown in Figure 23.5, as alternating layers of a conductor like aluminum foil, shown as the areas with lines across them, separated by layers of an insulator, like overhead transparencies or wax paper, shown as the blank areas in Figure 23.5.

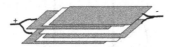

FIGURE 23.5   Capacitor made of alternating layers of conductors and insulators.

If every other layer of aluminum foil is connected on one end, the layers of foil will create a capacitor. If the positive side of a battery is connected to the left wire and the negative side, to right side the capacitor will store charge on the separated sheets of foil. Depending on the number of layers, area of the layers, and dielectric constant of the insulator between the conducting layers, the capacitance can be adjusted to a specific value of Farads. So, in a circuit, a capacitor looks more like the ones shown in Figure 23.6.

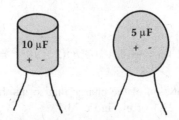

FIGURE 23.6   Sketch of common capacitors found in a circuit.

Sometimes the way they are connected, in terms of positive and negative, is important, and sometimes it doesn't matter. Look for the + and − on the capacitor to check if the direction it is installed in a circuit is critical.

23.4 CAPACITORS IN CIRCUITS

Thus far, only single capacitors across a voltage have been discussed. In most electrical circuits there are several capacitors connected in series, parallel, and a combination of the two, just like the case for resistors. In electrical circuit diagrams, the symbol for a capacitor is shown in Figure 23.7.

FIGURE 23.7   Circuit diagram symbol of a capacitor.

This symbol represents the two parallel plates of the simplest capacitor.

Kirchhoff's Laws are still the fundamental concepts employed to predict and understand the way capacitors behave in a circuit. When a circuit is connected, as shown in Figure 23.8, an electric field in the wires will pull negative charge of the positive plate of the capacitor, leaving it charged $(+q)$, and it will push negative charge onto the negative plate of the capacitor, leaving it charged $(-q)$.

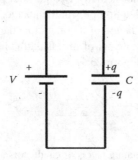

FIGURE 23.8   Circuit diagram of a battery and capacitor.

The time it takes to move this charge is known as the transient time of the circuit and is the focus of the next chapter, where the resistance and capacitance of the circuit are important in defining the time it takes to move the charge onto and off the plates. In the analysis of this chapter, the focus will be on what happens after the charge is distributed and the circuit is in a new equilibrium situation, often referred to as the steady state. In this steady-state arrangement, conservation of charge and Kirchhoff's Voltage Law (KVL) are the key in understanding how the charge is distributed on a circuit consisting of several capacitors.

23.4.1 CAPACITORS IN SERIES

In the series circuit of capacitors shown in Figure 23.9, when the circuit is connected the battery pulls negative charge off plate 1 of capacitor 1, C_1, leaving it positively charged, and it also pushes the same quantity of negative charge onto plate 4 of capacitor 2, C_2.

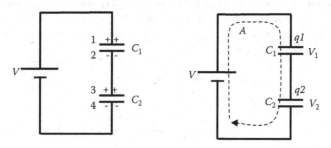

FIGURE 23.9 Series circuit of capacitors.

The excess positive charge on plate 1 attracts negative charge from plate 3 of capacitor 2, and the charge is balanced on the series capacitors. In summary, capacitors in series must have the same charge on their plates. Therefore, the voltage supplied by the battery is distributed so that the quantity of displaced charge is the same.

Applying KVL to the circuit of two capacitors in series in Figure 23.10, the total voltage supplied in loop A must equal the total voltage dissipated so

$$V = V_1 + V_2$$
$$V = \frac{q_1}{C_1} + \frac{q_2}{C_2}$$

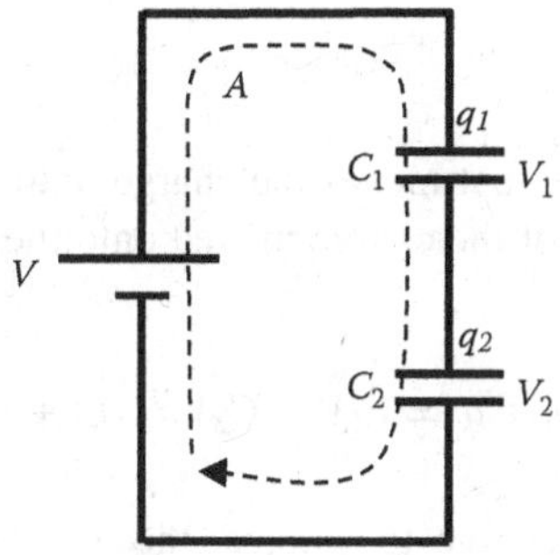

FIGURE 23.10 KVL for a series circuit of capacitors.

Since the charge must be the same on the plates of capacitors in series $q_1 = q_2 = q$, so the q can be factored out leaving

$$V = q \left(\frac{1}{C_1} + \frac{1}{C_2} \right)$$

Since capacitance is the ratio of charge to voltage, the equivalent capacitance of capacitors in series, Ceq, can be expressed in equation (23.6) as

$$\frac{1}{C_{eq}} = \frac{1}{C_1} + \frac{1}{C_2} = \sum_i \frac{1}{C_i} \qquad (23.6)$$

23.4.2 Capacitors in Parallel

In the parallel circuit of capacitors shown in Figure 23.11, when the circuit is connected, the battery pulls negative charge off plate 1 of capacitor 1, C_1, and off plate 3 of capacitor 2 leaving both plates positively charged, and it also pushes negative charge onto plate 2 of capacitor 1, C_1, and onto plate 4 of capacitor 2, C_2.

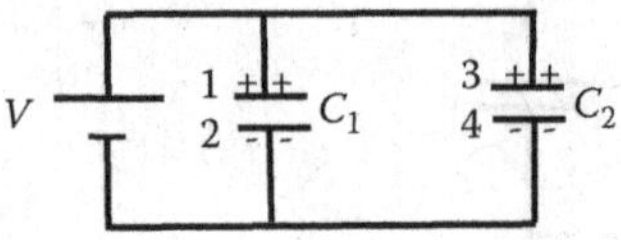

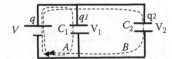

FIGURE 23.11 Parallel circuit of capacitors.

Applying KVL twice to the circuit of two capacitors in parallel produces two loops in Figure 23.12. The total voltage supplied in each of the loops, A and B, must equal the total voltage dissipated, so

$$V = V_1 \text{ and } V = V_2$$
$$V = \frac{q_1}{C_1} \text{ and } V = \frac{q_2}{C_2}$$

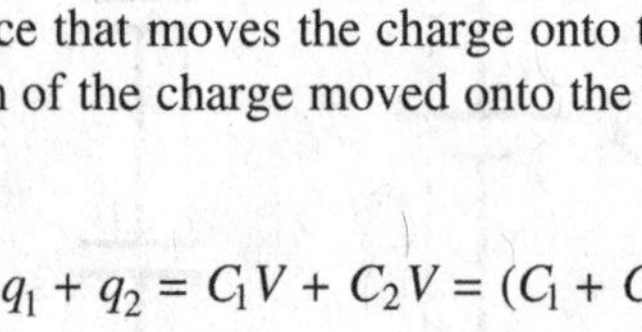

FIGURE 23.12 KVL for a parallel circuit of capacitors.

Since the battery is the only device that moves the charge onto the plates, the total charge moved by the battery, q, must be the sum of the charge moved onto the plates of capacitors in parallel, so $q = q_1 + q_2$,

$$q = q_1 + q_2 = C_1 V + C_2 V = (C_1 + C_2)V$$

Since capacitance is the ratio of charge to voltage, the equivalent capacitance of capacitors in parallel, Ceq, can be expressed in equation (23.7) as

$$C_{eq} = C_1 + C_2 = \sum_i C_i \qquad (23.7)$$

Unfortunately, the equations for the equivalent capacitance are opposite from the equations for equivalent resistance, but that is because capacitance is defined as the ratio of charge per voltage, $C = \frac{q}{V}$, and resistance is the ratio of voltage per current, $R = \frac{V}{I}$.

23.4.3 CAPACITOR CIRCUIT ANALYSIS EXAMPLE

Example 1

Find the voltage across and charge stored on each of the capacitors in Figure 23.13 if the capacitance of the capacitors is: $C_1 = 4 \ \mu F$, $C_2 = 6 \ \mu F$, and $C_3 = 12 \ \mu F$ and the voltage of the battery is $V = 9 \ V$.

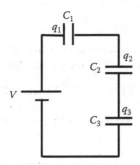

FIGURE 23.13 Electrical circuit for example 1.

Solution

Step 1. Label the charge stored on each capacitor in Figure 23.13.

Step 2. Find the equivalent capacitance of the series capacitors using equation (23.6).

$$\frac{1}{C_{eq}} = \frac{1}{C_1} + \frac{1}{C_2} + \frac{1}{C_3} = \frac{1}{4} + \frac{1}{6} + \frac{1}{12} = \frac{3}{12} + \frac{2}{12} + \frac{1}{12} = \frac{6}{12} = \frac{1}{2} \text{ so } C_{eq} = 2 \ \mu F$$

Step 3. Find the charge moved by the battery with equation (23.1), which is the definition of capacitance, $C = \frac{q}{V}$.

$$\text{So, } q = C_{eq} V = (2\mu F)(9 \ V) = 18 \ \mu C = 18 \times 10^{-6} \ C$$

This is the same charge for each of the capacitors since they are in series, so

$$q_1 = q_2 = q_3 = 18 \ \mu C = 18 \times 10^{-6} \ C$$

Step 4. Find the voltage across each capacitor with equation (23.1) solved for V, $V = \frac{q}{C}$, so

$$V_1 = \frac{q_1}{C_1} = \frac{18 \ \mu C}{4 \ \mu F} = 4.5 \ V, \quad V_2 = \frac{q_2}{C_2} = \frac{18 \ \mu C}{6 \ \mu F} = 3.0 \ V, \quad \text{and } V_3 = \frac{q_3}{C_3} = \frac{18 \ \mu C}{12 \ \mu F} = 1.5 \ V$$

Notice that $V_1 + V_2 + V_3 = 4.5 \ V + 3.0 \ V + 1.5 \ V = 9 \ V$

Example 2

Find the voltage across and charge stored on each of the capacitors in Figure 23.14 and the total charge moved by the battery, if the capacitance of the capacitors are: $C_1 = 4 \ \mu F$, $C_2 = 6 \ \mu F$, and $C_3 = 12 \ \mu F$ and the voltage of the battery is $V = 9 \ V$.

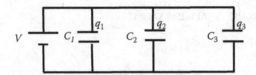

FIGURE 23.14 Electrical circuit for example 2.

Solution

Step 1. Label the charge stored on each capacitor in Figure 23.14.

Step 2. Find the equivalent capacitance of the parallel capacitors using equation (23.7).

$$C_{eq} = C_1 + C_2 + C_3 = 4 \ \mu F + 6 \ \mu F + 12 \ \mu F = 22 \ \mu F \quad \text{so} \quad C_{eq} = 22 \ \mu F$$

Step 3. Find the charge moved by the battery with equation (23.1), which is the definition of capacitance, $C = \frac{q}{V}$.

$$\text{So,} \quad q = C_{eq}V = (22 \ \mu F)(9 \ V) = 198 \ \mu C = 198 \times 10^{-6} \ C$$

Step 4. Since the capacitors are arranged in parallel across the battery, the voltage across each capacitor is the same as that of the battery, so $V = V_1 = V_2 = V_3$. With this information, the charge on each capacitor can be found,

$$q_1 = C_1 V_1 = (4 \ \mu F)(9 \ V) = 36 \ \mu C,$$
$$q_2 = C_2 V_2 = (6 \ \mu F)(9 \ V) = 54 \ \mu C, \quad \text{and}$$
$$q_3 = C_3 V_3 = (12 \ \mu F)(9 \ V) = 108 \ \mu C$$

Notice that $q_1 + q_2 + q_3 = 36 \ \mu C + 54 \ \mu C + 108 \ \mu C = 198 \ \mu C$

Example 3

Find the charge stored on and voltage across each of the capacitors in Figure 23.15 and the total charge moved by the battery, if the capacitance of the capacitors is: $C_1 = 4 \ \mu F$, $C_2 = 6 \ \mu F$, and $C_3 = 12 \ \mu F$ and the voltage of the battery is $V = 9 \ V$.

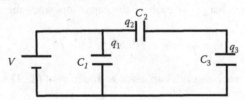

FIGURE 23.15 Electrical circuit for example 3.

Step 2. Find the equivalent capacitance of the C_2 and C_3, which are in series using equation (23.6).

$$\frac{1}{C_{23}} = \frac{1}{C_2} + \frac{1}{C_3} = \frac{1}{6} + \frac{1}{12} = \frac{2}{12} + \frac{1}{12} = \frac{3}{12} = \frac{1}{4} \quad \text{so} \quad C_{23} = 4 \ \mu F.$$

Find the equivalent capacitance of the total circuit capacitors using equation (23.7).

$$C_{eq} = C_1 + C_{23} = 4 \ \mu F + 4 \ \mu F = 8 \ \mu F \quad \text{so} \quad C_{eq} = 8 \ \mu F$$

Step 3. Find the charge moved by the battery with equation (23.1), which is the definition of capacitance, $C = \frac{q}{V}$.

So, $q = C_{eq}V = (8\ \mu F)(9\ V) = 72\ \mu C = 72 \times 10^{-6}\ C$.

Step 4. Since capacitor 1 is in parallel with the combination of capacitors 2 and 3 and with the battery, the voltage across C_1 and C_{23} is the same as that of the battery, so $V = V_1 = V_{23}$. With this information, the charge on these capacitors can be found,

$$q_1 = C_1 V_1 = (4\ \mu F)(9\ V) = 36\ \mu C, \quad \text{and} \quad q_{23} = C_{23}V_{23} = (4\ \mu F)(9\ V) = 36\ \mu C$$

Notice that $q_1 + q_{23} = 36\ \mu C + 36\ \mu C = 72\ \mu C$

The last step is finding the voltage and charge across each of capacitor 2 and capacitor 3. Since these capacitors are in series, they each have 36 μC of charge on them, so the voltages are: $V_2 = \frac{q_2}{C_2} = \frac{36\ \mu C}{6\ \mu F} = 6.0\ V$ and $V_3 = \frac{q_3}{C_3} = \frac{36\ \mu C}{12\ \mu F} = 3.0\ V$, which together add to the 9 V across the battery.

From these examples, it is clear that Kirchhoff's Voltage Law, which is conservation of energy per charge, holds for capacitors as well as resistors and is a key concept used in analyzing electrical circuits.

23.5   ANSWER TO CHAPTER QUESTION

As you many have figured out, the large electrical shock delivered by a portable defibrillator comes from charging and discharging a capacitor bank. As described in Figure 23.16, the first step is to close switch 1, which allows the capacitor C to be charged by the voltage supply (V).

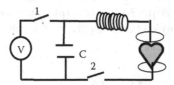

FIGURE 23.16   Simple defibrillator.

When the prescribed amount of charge is pushed onto the opposing plates of the capacitor, switch 1 is open and switch 2 is closed to deliver a large pulse of electrical energy to the heart.

23.6   QUESTIONS AND PROBLEMS

23.6.1   MULTIPLE CHOICE QUESTIONS

1. A parallel-capacitor is made of two square plates with an area, A, that are positioned parallel to each other at distance, d, apart, with air between them. A voltage supply is connected across the plates, set at a voltage of V. Please see Figure 23.1 for a depiction of this situation. If the supply remains connected across the plates and the distance between the plates is decreased, will the charge stored on the plates increase, decrease, or remain the same?
   A.   increase
   B.   decrease
   C.   remain the same

2. Capacitors $C_1 = 1$ μF, $C_2 = 2$ μF, $C_3 = 3$ μF, and a 6 V battery are all connected in *series* with each other. Which capacitor has the highest voltage across it?
   A. C_1
   B. C_2
   C. C_3
   D. The voltage across all the capacitors is the same.
3. Capacitors $C_1 = 1$ μF, $C_2 = 2$ μF, $C_3 = 3$ μF, and a 6 V battery are all connected in series with each other. Which capacitor stores the greatest amount of charge?
   A. C_1
   B. C_2
   C. C_3
   D. The capacitors all store the same amount of charge.
4. Capacitors $C_1 = 1$ μF, $C_2 = 2$ μF, $C_3 = 3$ μF, and a 6 V battery are all connected in parallel with each other. Which capacitor has the highest voltage across it?
   A. C_1
   B. C_2
   C. C_3
   D. The voltage across all the capacitors is the same.
5. Capacitors $C_1 = 1$ μF, $C_2 = 2$ μF, $C_3 = 3$ μF, and a 6 V battery are all connected in parallel with each other. Which capacitor stores the greatest amount of charge?
   A. C_1
   B. C_2
   C. C_3
   D. The capacitors all store the same amount of charge.

The following description and Figure 23.17 are for multiple choice questions 6–10 and problems 9 and 10. Four capacitors A, B, C, and D are connected, as demonstrated in Figure 23.15. The capacitance of capacitors A, B, C, and D is 10 μF, 20 μF, 30 μF, and 40 μF, respectively. These capacitors are charged by connecting the combination across a power supply, with a voltage of 4 kV.

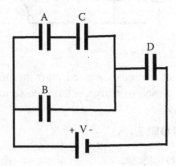

FIGURE 23.17   Diagram for multiple choice question 6–10 and problems 9 and 10.

6. Which capacitor A, B, C, or D stores the greatest amount of charge on its plates?
   A. A
   B. B
   C. C
   D. D
   E. All charges are the same.

7. Is the charge stored by capacitor A greater than, less than, or equal to the charge stored by capacitor C?
   A. greater than
   B. less than
   C. equal to
8. Is the charge stored by capacitor A greater than, less than, or equal to the charge stored by capacitor B?
   A. greater than
   B. less than
   C. equal to
9. Is the voltage across capacitor A greater than, less than, or equal to the voltage across capacitor B?
   A. greater than
   B. less than
   C. equal to
10. Is the voltage across capacitor A greater than, less than, or equal to the voltage across capacitor C?
    A. greater than
    B. less than
    C. equal to

23.6.2 PROBLEMS

1. Find the charge on the plates of a 10 μF capacitor with 5 kV across it.
2. A parallel plate capacitor, similar to the one depicted in Figure 23.5, is made with two pieces of aluminum foil, a good conductor, and one layer of waxed paper. The capacitor has a capacitance of 2 nF, and the pieces of waxed paper are separated by the 0.5 mm thickness of the waxed paper. Compute the area of the pieces of aluminum foil and, assuming the foil is a square, find a length of a side of the foil.
3. Compute the energy stored in a capacitor with a capacitance of 5 μF and a voltage of 3 kV across the plates.
4. Compute the voltage across the parallel plates of a capacitor that is separated by 4 cm if the electric field between the plates has a magnitude of 5×10^5 N/C.
5. Find the charge on a capacitor consisting of two metal disks of diameter 6.00 mm separated by a 5 nm thick layer of titanium dioxide when the capacitor is connected across a 3 V battery.
6. Consider a parallel plate capacitor, whose plate separation is adjustable. The plates are in the shapes of disks of diameter 20.0 cm. The plate separation is 1.5 mm. There is nothing but air between the plates. A person connects a 6.15 V battery across the capacitor. After the capacitor is charged, the battery is disconnected from the capacitor. Then, the plate separation is increased to 2.00 mm. Find the voltage across the capacitor at the new plate separation.
7. Find the voltage across and charge stored on capacitor 3 (C_1) in Figure 23.18, if the capacitance of the capacitors are: $C_1 = 2\ \mu F$, $C_2 = 3\ \mu F$, and $C_3 = 5\ \mu F$ and the voltage of the battery is $V = 10$ V.

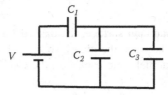

FIGURE 23.18 Diagram for problems 7 and 8.

8. Find the voltage across and charge stored on capacitor 3 (C_3) in Figure 23.18, if the capacitance of the capacitors is: $C_1 = 2\ \mu F$, $C_2 = 3\ \mu F$, and $C_3 = 5\ \mu F$ and the voltage of the battery is $V = 10$ V.

9. In Figure 23.17, 4 capacitors, A, B, C, and D, are connected as demonstrated. The capacitance of capacitors A, B, C, and D is 10 μF, 20 μF, 30 μF, and 40 μF, respectively. These capacitors are charged by connecting the combination across a power supply, with a voltage of 4 kV. For this circuit, compute the equivalent capacitance and the total charge moved by the power supply.

10. In Figure 23.17, 4 capacitors, A, B, C, and D, are connected as demonstrated. The capacitance of capacitors A, B, C, and D is 10 μF, 20 μF, 30 μF, and 40 μF, respectively. These capacitors are charged by connecting the combination across a power supply, with a voltage of 4 kV. For this circuit, compute the magnitude of the charge stored on the left plate of capacitor **B**.

APPENDIX GAUSS' LAW FOR A PARALLEL PLATE CAPACITOR

In Chapter 5, Gauss's Law was introduced as a geometric way to understand the electric field, and it was expressed as equation (5.4):

$$E\ A_p = \frac{1}{\varepsilon_o} \sum Q_{enc} \tag{5.4}$$

where the permittivity of free space has a value of $\varepsilon_0 = 8.85 \times 10^{-12}\ \frac{C^2}{N \cdot m^2}$, E, is the electric field at a region in space, A_p is the surface area of a shape that encloses the field such that the surface and the field are everyplace perpendicular and $\sum Q_{enc}$ is the net charge enclosed by the shape.

Consider the positive plate of a charged parallel plate capacitor in Figure 23.19.

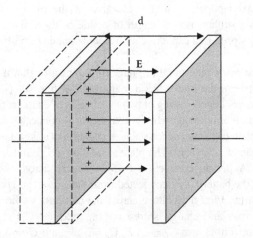

FIGURE 23.19 Gaussian surface and electric field of a parallel plate capacitor.

The charge enclosed by the Gaussian surface, the dashed line, is q. The Gaussian surface is perpendicular to the electric field between the plates, so $A_p = A = $ Area of the plate.

So, by Gauss's Law for this parallel plate capacitor, the electric field between the plates is:

$$E = \frac{q}{\varepsilon_o A_p}$$

Using the relationship between the voltage across the plates and the electric field between the plates presented in equation (23.5), the voltage across a parallel plate capacitor is:

$$V = \frac{U}{q} = Ed = \frac{qd}{\varepsilon_o A_p}$$

Because capacitance is defined in equation (23.1) as $C = \frac{q}{V}$, the capacitance of a parallel plate capacitor can be found through Gauss's Law as:

$$C = \frac{q}{V} = \frac{q}{V \frac{qd}{\varepsilon_o A_p}} = \varepsilon_o \frac{A}{d}$$

If the multiplication of the electric field by a dielectric put between the plates, equation (3.2) results:

$$C = \kappa \varepsilon_o \frac{A}{d}$$

This expression relates the geometry of the capacitor to its electrical capacitance in the way the geometry of a bucket is related to its capacitance to store water. It makes sense that if the plates are made larger, the capacitance increases, and if the plates are brought closer together, the capacitance goes up. Also, if the geometry of the capacitor is changed, the equation will change, since this derivation was done for the geometry of a parallel plate capacitor.

24 RC Circuits

24.1 INTRODUCTION

The time dependence of an electrical circuit that contains both a capacitor and a resistor, called and RC circuit, can be used to help understand how the signal travels down the axon of a nerve cell. This comparison requires the synthesis of ideas from all three chapters on electrical circuits and is the end goal of this chapter on RC circuit analysis. The chapter begins with the application of Kirchhoff's Laws to the analysis of the charging of an initially grounded capacitor. Then, these same techniques are applied to the discharging of a charged capacitor in an RC circuit. Together the analysis is applied to answer the chapter question, which is about the analysis of the signal down the axon of a nerve cell.

24.2 CHAPTER QUESTION

What electrical circuit is the best model for understanding the signals that travel down the axon of the neuron? This question will be answered at the end of this chapter and provides an opportunity to synthesize the material from all three chapters on electrical circuit analysis.

24.3 RC CIRCUITS

24.3.1 RC CIRCUIT – INITIALLY GROUNDED CAPACITOR

The first step to understand the electrical signal across a biological cell membrane is to investigate the RC circuit, which is a circuit constructed from a battery, a switch, a capacitor, and resistors, as depicted in Figure 24.1. With the switch set to position 2 in Figure 24.1, the capacitor will be grounded so that the voltage across and charge stored by the capacitor is zero. This is the initial condition of the circuit.

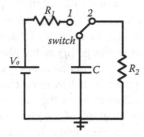

FIGURE 24.1 Initially grounded RC circuit.

24.3.2 RC CIRCUIT – CHARGING

If at a time of zero seconds ($t = 0$ s) the switch is moved from position 2 to position 1, a current (I) will begin to flow through resistor R_1 and charge, q, will begin to collect on the plates of the capacitor, C, as depicted in Figure 24.2.

The charge will build up on the plates of the capacitor, C, until the voltage across the capacitor matches the voltage, V_o, across the battery. This is the process known as charging the capacitor.

DOI: 10.1201/9781003308072-25

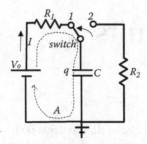

FIGURE 24.2 Charging the capacitor in an RC circuit.

The analysis of the circuit begins with applying Kirchhoff's Voltage Law (KVL) around loop A sketched in Figure 24.2, where the battery provides a voltage V_o and resistor 1 and the capacitor dissipate this voltage.

$$V_0 = V_{R1} + V_C$$

By Ohm's Law and the definition of capacitance,

$$V_0 - IR_1 - \frac{q}{C} = 0$$

$$IR_1 = V_0 - \frac{q}{C}$$

$$I = \frac{V_0}{R_1} - \frac{q}{R_1 C}$$

Since, the current is equal to the time rate of change of the charge on the capacitor, $I = \Delta q/\Delta t$, to get:

$$\frac{\Delta q}{\Delta t} = \frac{V_0}{R_1} - \frac{q}{R_1 C}.$$

This equation states that the rate of change of the charge on the plates, $\frac{\Delta q}{\Delta t}$, of the capacitor, depends on the quantity of charge, q, on the capacitor. Systems that follow this type of a relationship in which the rate of change of a quantity depends on the amount of that same quantity are exponential.

In this case, the quantity of the excess charge (q) on the plates starts at zero, increases rapidly, and then slows down as the quantity of charge on the plates grows. The reason for this slowing of the rate of change of charge on the plates is that as more charge is deposited on the plates, the repulsive electrostatic force on the next quantity of the same sign of charge that is being pushed onto or pulled off the plates increases. A graph of the charge on the plates of the capacitor as a function of time, starting at the instant the switch moved from position 1 to position 2 in Figure 24.2, is given in Figure 24.3.

The quantity of charge (q) as a function of time that follows from the equation of the time rate of change of charge and that fits the curve of a charging capacitor in a circuit shown in Figure 24.2 is given by equation (24.1) as

$$q = q_o\left(1 - e^{-\frac{t}{\tau}}\right) \qquad\qquad (24.1)$$

where $q_o = CV_o$ and $\tau = R_1 C$. The quantity τ is the time constant for this circuit. Since the Ohm is just the ratio of volts to amps, by Ohm's Law, and the Farad is the ratio of coulombs per volt, by

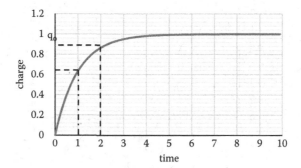

FIGURE 24.3 Charge stored on the plates of the charging capacitor vs. time.

the definition of capacitance, the unit generated by the product of units for resistance (Ω) and capacitance (F) is the second,

$$[\Omega F] = \left[\left(\frac{V}{A}\right)\left(\frac{C}{V}\right)\right] = \frac{C}{A} = \frac{C}{C/s} = s.$$

The mathematical proof of this exponential solution requires integral calculus, but the understanding of it is an important interrelationship that occurs often in nature, and that is why the exponential and its inverse, the natural log, provides a mathematical model for many systems.

If the time in equation (24.1) is set to one time constant, $t = \tau$, the value of the charge is:

$$q = q_o\left(1 - e^{-\frac{\tau}{\tau}}\right) = 0.63\ q_o.$$

So, at a time of one time constant, the value of q reaches 63% of its maximum value. If the time in equation (24.1) is set to two time constants, $t = 2\tau$, the value of the charge is:

$$q = q_o\left(1 - e^{-\frac{2\tau}{\tau}}\right) = 0.86\ q_o.$$

So, at a time of two time constants, the value of q reaches 86% of its maximum value. These two points in time of one and two time constants are important markers in the growth of the charge across the capacitor plates or any other quantity changing with this type of behavior.

The expression for the voltage across a charging capacitor follows a similar expression given in equation (24.2), with V_o equal to the voltage of the battery.

$$V = V_o\left(1 - e^{-\frac{t}{\tau}}\right) \tag{24.2}$$

A graph of the voltage across the plates of the capacitor as a function of time, starting at the instant the switch moved from position 1 to position 2 in Figure 24.2, is given in Figure 24.4.

The same points of one and two time constants hold for this graph with the same values as those for charge. This should be the case, since the charge stored on the plates of a capacitor, q, and the voltage, V, across the plates of a capacitor are related by the definition of capacitance, $C = q/V$. So, the charge across the plates is just the product of a constant value of the capacitance and the voltage across the plates, $q = CV$, so the graphs and equations should have the same time dependence.

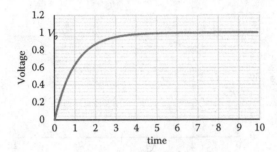

FIGURE 24.4   Voltage across the plates of the charging capacitor vs. time.

On the other hand, the current, I, through R_1 will start with a maximum value and decrease as time progresses, since it is more difficult to move charge onto the plates of the capacitor as charge builds. Therefore, the current I through the charging resistor, R_1, is given by equation (24.3) as

$$I = I_o \left(e^{-\frac{t}{\tau}} \right). \tag{24.3}$$

where I_o is the initial current through resistor 1, when the switch is moved from position 1 to position 2 in Figure 24.2. This initial current will have a value of $I_o = V_o/R_1$ by Ohm's Law. A graph of the current through the charging resistor, R_1, as a function of time, starting at the instant the switch moved from position 2 to position 1 in Figure 24.2, is given in Figure 24.5.

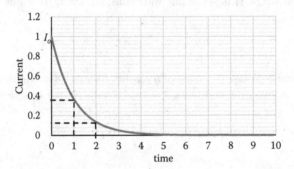

FIGURE 24.5   Current through the charging resistor vs. time.

If the time in equation (24.3) is set to one time constant, $t = \tau$, the value of the current is:

$$I = I_o \left(e^{-\frac{\tau}{\tau}} \right) = 0.37 \ q_o.$$

So, at a time of one time-constant, the value of I reaches 37% of its initial value. If the time in equation (24.3) is set to two time-constants, $t = 2\tau$, the value of the charge is:

$$I = I_o \left(e^{-\frac{2\tau}{\tau}} \right) = 0.135 \ I_o.$$

So, at a time of two time constants the value of q reaches 13.5% of its maximum value. These two points in time of one and two time constants are important markers in the decrease of the current through the charging resistor or any other quantity changing with similar behavior.

24.3.3   RC Circuit – Discharging

Assuming the capacitor, C, in Figure 24.6 is fully charged so that the voltage across it is equal to the voltage across the battery, Vo, and it has a quantity of excess charge of $q_o = CV_o$ on its plates.

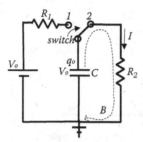

FIGURE 24.6   Discharging the capacitor in an RC circuit.

The clock is reset and at a new start time of zero seconds ($t = 0$ s) the switch is moved from position 1 to position 2, a current (I) will begin to flow through resistor R_2 and charge, q, will begin to move off the plates of the capacitor, C, as depicted in Figure 24.6. The charge flow off the plates of the capacitor, C, until the voltage across the capacitor is equal to the voltage of the ground, which is commonly set to zero. This process is known as *discharging a capacitor*.

The analysis of the circuit begins with applying Kirchhoff's Voltage Law (KVL) around loop B sketched in Figure 24.6, where there is no voltage source and just the capacitor and resistor 2.

$$0 = V_c + V_{R2}$$

By Ohm's Law and the definition of capacitance,

$$\frac{q}{C} = -IR_2$$

$$-I = \frac{q}{R_2 C}$$

Since the current is equal to the time rate of change of the charge on the capacitor, the quantity $I = \Delta q/\Delta t$, to get:

$$\frac{-\Delta q}{\Delta t} = \frac{q}{R_2 C}$$

This equation is rearranged resulting in: $\frac{\Delta q}{\Delta t} = \frac{-q}{R_2 C}$. As with the equation for the charging capacitor, this equation states that the rate of change of the charge, $\frac{\Delta q}{\Delta t}$, flowing off the plates of the capacitor, depends on the quantity of charge, q, on the capacitor. As stated before, systems that follow this type of a relationship in which the rate of change of a quantity depends on the amount of that same quantity are exponential. Unlike the case for the charging of the capacitor, there is no additional term in the solution, because there is no voltage supply in the circuit that is being analyzed.

In this case, the quantity of the excess charge (q) on the plates starts at q_o, decreases rapidly, and then slows down as the quantity of charge on the plates decreases. The reason for this slowing of the rate of change of charge on the plates is that as more charge is discharged to ground there is a decrease in the remaining charge that is the source of the repulsive electrostatic force on the next quantity of charge of the same sign to be pushed off the plates to ground. A graph of the charge on the plates of the capacitor as a function of time, starting at the instant the switch moved from position 1 to position 2 in Figure 24.6, is given in Figure 24.7.

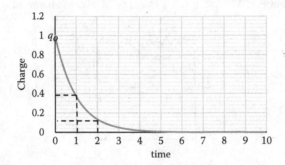

FIGURE 24.7   Charge stored on the plates of the discharging capacitor vs. time.

The quantity of charge (q) as a function of time that follows from the equation of the time rate of change of charge and that fits the curve of a discharging capacitor in a circuit shown in Figure 24.6 is given by equation (24.4) as

$$q = q_o\left(e^{-\frac{t}{\tau}}\right) \tag{24.4}$$

where $q_o = CV_o$ and $\tau = R_2C$. The quantity τ is the time constant for this discharging circuit.

If the time in equation (24.4) is set to one time constant, t = τ, the value of the charge is:

$$q = q_o\left(e^{-\frac{\tau}{\tau}}\right) = 0.37\ q_o.$$

So, at a time of one time constant the value of q reaches 37% of its maximum value. If the time in equation (24.4) is set to two time constants, $t = 2\tau$, the value of the charge is:

$$q = q_o\left(e^{-\frac{2\tau}{\tau}}\right) = 0.135\ q_o.$$

So, at a time of two time constants, the value of q reaches 13.5% of its maximum value. These two points in time of one and two time constants are important markers in the growth of the charge across the capacitor plates or any other quantity changing with this type of behavior.

The expression for the voltage across the discharging capacitor follows a similar expression given in equation (24.5), with V_o equal to the voltage of the battery.

$$V = V_o\left(e^{-\frac{t}{\tau}}\right) \tag{24.5}$$

A graph of the voltage across the on the plates of the capacitor as a function of time, starting at the instant the switch moved from position 1 to position 2 in Figure 24.6, is given in Figure 24.8.

The same points of one and two time constants hold for this graph with the same values as those for the voltage across as the charge for a discharging capacitor.

Unlike the charging capacitor, the current through the discharging resistor, R_2, follows the same behavior as the charge and voltage. In that, the current, I, through R_2 starts with a maximum value of I_o and decrease as time progresses. This is the case, since the quantity of charge on the plates of

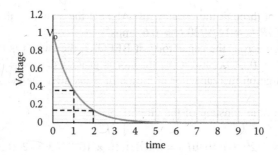

FIGURE 24.8 Voltage across the plates of the charging capacitor vs. time.

the capacitor decreases with time, and it is this excess charge on the plates that provides the force needed to push the charge through R_2 to the ground. Therefore, the current I through the discharging resistor, R_2, is given by equation (24.6) as

$$I = I_o\left(e^{-\frac{t}{\tau}}\right). \tag{24.6}$$

where I_o is the initial current through resistor 2, when the switch is moved from position 1 to position 2 in Figure 24.2. This initial current will have a value of $I_o = V_o/R_2$ by Ohm's Law. A graph of the current through the discharging resistor, R_2, as a function of time, starting at the instant the switch moved from position 1 to position 2 in Figure 24.6, is given in Figure 24.9.

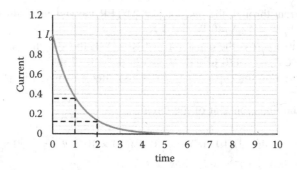

FIGURE 24.9 Current through the discharging resistor vs. time.

The same points of one and two time constants hold for this graph with the same values for current as those for the charge and voltage for a discharging capacitor.

24.4 RC CIRCUIT EXAMPLES

24.4.1 CHARGING EXAMPLE

Considering the circuit in Figure 24.2 with a battery voltage of $V_o = 12$ V, the capacitance of the capacitor is $C = 10$ μF, and the resistance of the charging resistor is $R_1 = 20$ Ω. The capacitor is initially grounded, as shown in Figure 24.1, and the switch moved from position 2 to position 1 at $t = 0$ s.

a. What is the current through the resistor just BEFORE the switch is thrown?
   Solution: $I = 0\ A$

b. What is the current through the resistor just AFTER the switch is thrown?
   Solution: $I_o = V_o/R$ $I_o = 12$ V/20 $\Omega = 0.6$ amps
c. What is the charge across the capacitor just BEFORE the switch is thrown?
   Solution: $q = CV = 0$ C, because the capacitor is grounded.
d. Compute the charge on the capacitor at a time of 3 ms after the switch is thrown.
   Solution:
   Step 1. Find the time constant of the circuit:

$$\tau = R_l C = (20\ \Omega)(10\ \mu F) = (20\ \Omega)(10 \times 10^{-6}\ F) = (200 \times 10^{-6}\ s)$$
$$\tau = 0.2 \times 10^{-3}s = 0.2\ ms$$

   Step 2. Find the maximum charge: $q_o = CV_o = (10\ mF)(12\ V) = 120\ mC$
   Step 3. Find the charge at the time $t = 3$ ms:

$$q = q_o\left(1 - e^{-\frac{t}{\tau}}\right) = 120\ mC\left(1 - e^{-\frac{0.3\ ms}{0.2\ ms}}\right) = 93.2\ mC$$

24.4.2 DISCHARGING EXAMPLE

Consider the circuit in Figure 24.6 with a battery voltage of $V_o = 12$ V, the capacitance of the capacitor is $C = 10$ μF, and the resistance of the dissipating resistor is $R_2 = 40\ \Omega$. The switch in the circuit was in position 1 for a long time. At time $t = 0$, the switch is moved to position 2.

a. What is the current through the resistor just AFTER the switch is thrown?
   Solution: $I_o = V_o/R = 12$ V/40 W $= 0.3$ A
b. What is the charge across the capacitor just BEFORE the switch is thrown?
   Solution: $q_o = CV_o$ $q_o = (10\ \mu F)(12\ V) = 120\ \mu C = 0.12$ mC
c. What is the charge on the capacitor ($t = 0.3$ ms) after the switch is thrown?
   Solution:
   Step 1. Find the time constant of the circuit:

$$\tau = R_2 C = (40\ \Omega)(10\ \mu F) = (40\ \Omega)(10 \times 10^{-6}\ F) = (400 \times 10^{-6}\ s)$$
$$\tau = \tau = 0.4 \times 10^{-3}s = 0.4\ ms$$

   Step 2. Find the maximum charge: $q_o = CV_o = (10\ mF)(12\ V) = 120\ mC$
   Step 3. Find the charge at the time $t = 3$ ms:

$$q = q_o\left(e^{-\frac{t}{\tau}}\right) = 120\ mC\left(e^{-\frac{0.3\ ms}{0.4\ ms}}\right) = 56.7\ mC$$

d. At what time after the switch has moved from position 1 to position 2 is the voltage across the capacitor equal to 4 V?
   Solution:
   Step 1. Find the time constant of the circuit, which was found in part d as $\tau = 0.4 \times 10^{-3}$ s $= 0.4$ ms
   Step 2. Find the maximum voltage, which was given as: $V_o = 12$ V
   Step 3. Find the time at which voltage across the capacitor is $V = 6$ V:

Start with equation (24.5): $V = V_0\left(e^{-\frac{t}{\tau}}\right)$

Divide both sides by V_o: $\frac{V}{V_0} = \left(e^{-\frac{t}{\tau}}\right)$

Take the natural log of both sides: $ln\left(\frac{V}{V_0}\right) = ln\left(e^{-\frac{t}{\tau}}\right)$

The natural log is the inverse of the exponential, so $ln(e^x) = x$. Therefore:

$$ln\left(\frac{V}{V_0}\right) = \left(-\frac{t}{\tau}\right)$$

Solve for t:

$$t = (-\tau)ln\left(\frac{V}{V_0}\right)$$

Plug in the numbers to find the time at which the voltage across the capacitor is 6 V, when it was at 12 V at $t = 0$ s.

$$t = (-\tau)ln\left(\frac{V}{V_0}\right) = (-0.4\ ms)\left[ln\left(\frac{6\ V}{12\ V}\right)\right] = (-0.4\ ms)[-0.693] = 0.277\ ms$$

Notice that the natural log of a fraction is a negative number, so the resulting time is a positive value. Also notice that the time for the voltage to reach half its initial value is not τ, which is also the symbol used for half-life, but in a completely different context.

24.5   ANSWER TO CHAPTER QUESTION (SYTHESIS OPPORTUNITY 1)

Nerve cells, or neurons, transport electrical signals throughout the human body. In Figure 24.10, the cell has a nucleus inside the body of the cell, an axon that ends in dendrites that then connect to other cells.

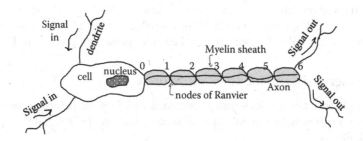

FIGURE 24.10   Sketch of the axon of a nerve cell.

These dendrites connect to the cell body of other cells, and the signal moves though the body. The cause from several neurological issues, such as multiple sclerosis (MS), is thought to be a degradation of the electrical signal as it travels down the axon of the neuron.

24.5.1   CIRCUIT MODEL OF THE AXON

To better understand the transport of the signal down the axon, researchers have developed electrical circuit models of the axon of a neuron. This analysis is based upon RC circuits and combinations of resistors, and an example of a circuit used to study the electrical properties of the axon of a nerve cell is given in Figure 24.11.

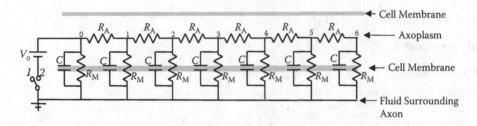

FIGURE 24.11 A circuit used to model a segment of an axon.

In this diagram, the axoplasm is the cellular material in the axon, and the gaps between the myelin sheaths are known as the nodes of Ranvier. The membrane of an axon, along with its myelin sheath, acts as a dielectric separating the axoplasm from the fluid surrounding the axon. Thus, each of the segments of an axon is a capacitor whose plates are the outer surface of the axoplasm, in contact with the inside surface of the axon membrane, and the inner surface of the fluid surrounding the axon, which is in contact with the outside surface of the axon membrane. The membrane, however, is not a perfect insulator. Some charge can flow through it from the axoplasm, through the membrane, to the fluid surrounding the axon. In this model, the resistors labeled R_M (the sub-M stands for *membrane*) are in parallel with the capacitor. The smaller axoplasm subdivisions are represented by resistors in series with each other, each having a resistance R_A, here the sub-A stands for *axoplasm*. This is a model of a 1 mm long segment of an axon from one node to the next, or an axon that is only 1 mm long. Compare this with axons in humans that extend all the way from the spinal column to the foot. In this model, the 1 mm axon segment is broken up into six smaller segments, labeled with numbers from 0 to 6 at the locations of the circuit that are known as nodes. The gaps between these numbers are the gaps in the myelin sheaths that are known as the nodes of Ranvier. The resistors and capacitors are used to model these six discrete steps in the 1 mm axon segment.

In this model, the fluid surrounding the axon is set to be at 0 V. Then, the voltage at node 0 in Figure 24.11, is the voltage difference between node 0 and the fluid surrounding the axon, which is the voltage across the corresponding membrane resistor R_M. The signal starts down the axon when the switch just below the voltage supply is switched from point 1 to point 2, in Figure 24.11.

24.5.2 Resistance and Capacitance of the Neuron

The resistance and capacitance values were determined based on the published properties[1-4] of a 1 mm segment of a representative unmyelinated axon of inner radius $r = 5$ μm $= 5 \times 10^{-6}$ m, as described in Figure 24.12.

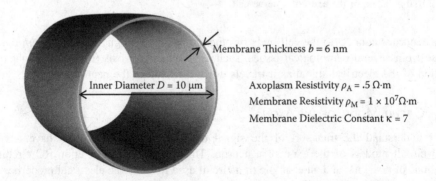

Membrane Thickness $b = 6$ nm

Inner Diameter $D = 10$ μm

Axoplasm Resistivity $\rho_A = .5$ Ω·m
Membrane Resistivity $\rho_M = 1 \times 10^7$ Ω·m
Membrane Dielectric Constant $\kappa = 7$

FIGURE 24.12 A cross-sectional view of a segment of an axon.

The interior of the axon is filled with a conducting material known as axoplasm. The axon is inside another tube, which contains a fluid that is a good conductor compared to both the axon membrane and the axoplasm. In this model, the fluid surrounding the axon is considered an ideal conductor. The 1 mm long segment of the axon is broken up into six $\frac{1}{6}$ mm segments. Six steps were chosen for this model. More steps would make a better approximation, but they are even more difficult to analyze.

The cross-sectional area of the axoplasm is,

$$A_A = \pi r^2 = \pi (5 \times 10^{-6} \text{ m})^2 = 7.854 \times 10^{-11} \text{ m}^2.$$

Plugging this area and the subdivision length of $L_S = \frac{1}{6}$ mm into equation (22.10), which is an expression for the resistance in terms of resistivity, results in a resistance of the axoplasm of

$$R_A = \rho_A \frac{L_S}{A_A} = .5\Omega \cdot \text{m} \frac{\frac{1}{6} \times 10^{-3} \text{m}}{7.854 \times 10^{-11} \text{m}^2} = 1,061,030 \ \Omega.$$

The membrane of an axon, along with its myelin sheath, acts as a dielectric separating the axoplasm from the fluid surrounding the axon. Thus, a segment of an axon is a capacitor whose plates are the outer surface of the axoplasm, which is in contact with the inside surface of the axon membrane, and the inner surface of the fluid surrounding the axon, and in contact with the outside surface of the axon membrane. For the capacitance of a 1/6 mm length of the axon, the area A_M we used for the resistance is the area of the capacitor plate and the thickness b of the membrane is the distance between the plates. Plugging into the expression of capacitance in equation (23.2) generates the capacitance of the membrane as,

$$C_M = \kappa \varepsilon_o \frac{A_M}{b} = 7 \left(8.85 \times 10^{-12} \frac{\text{F}}{\text{m}} \right) \frac{5.236 \times 10^{-9} \text{ m}^2}{6 \times 10^{-9} \text{ m}} = 5.4062 \times 10^{-11} \text{ F}.$$

The membrane, however, is not a perfect insulator. Some charge can flow through it, from the axoplasm, through the membrane, to the fluid surrounding the axon. This is modeled as a resistor in parallel with the capacitor. For the membrane resistor, the distance through the membrane is just the thickness of the membrane $b = 6$ nm; this is the length of the object whose resistance is calculated. To find the cross-sectional area, it is assumed each 1/6 mm segment of the tube at the bottom is unrolled and flattened so it has a height $b = 6$ nm, a width of $2\pi r$, and a length of 1/6 mm. Therefore, its cross-sectional area is:

$$A_M = (2\pi r) L_S = 2\pi (5 \times 10^{-6} \text{ m}) \frac{1}{6} \times 10^{-3} \text{ m} = 5.236 \times 10^{-9} \text{ m}^2$$

Plugging this and the length $b = 6$ nm, the thickness of the membrane, into equation (22.10) generates a resistance of the membrane of,

$$R_M = \rho_M \frac{b}{A_M} = 1 \times 10^7 \Omega \cdot \text{m} \frac{6 \times 10^{-9} \text{ m}}{5.236 \times 10^{-9} \text{ m}^2} = 11,459,129 \ \Omega.$$

Therefore, the values of resistance and capacitance for the model for a 1 mm segment of an unmyelinated axon are:

$$V_o = 110 \ mV, \quad R_A = 1.061 \times 10^6 \ \Omega, \quad R_M = 1.1459 \times 10^7 \ \Omega, \quad C_M = 5.4062 \times 10^{-11} \text{ F}$$

The fundamental function of the axon is the transmission of electrical signals along its length. An ion pump within the nerve cell prepares an axon for signal transmission by creating a potential difference between the axoplasm and the fluid surrounding the axon. As a signal arrives at one of the nodes of Ranvier, the electric potential of the axoplasm at that location increases. An increase in the electric potential of the axoplasm, above a threshold value, causes a change in the permeability of the axon membrane, thus flooding the axoplasm with positive ions from the fluid surrounding the axon, amplifying the increase in the electric potential enough for the signal to make it to the next node of Ranvier. The original increase in the electric potential triggers an action potential, amplifying the signal so that it can trigger another action potential at the next node of Ranvier. The circuit used to represent the segments of an axon in this section of the chapter only models the passive electrical characteristics, not this amplification from one node of Ranvier to the next. Thus, this circuit can also be considered to represent a short axon, on the order of a millimeter in length, with no amplification of the signal provided.

Since the fluid surrounding the axon is at an electric potential of 0 V, the electric potential at node 0 in the diagram above, is the electric potential difference between node 0 and the fluid surrounding the axon. This is just the voltage across the corresponding membrane resistor R_M.

24.5.3 PASSIVE RESISTIVE SIGNAL DEGRADATION

If the switch on the lower left of the circuit below is closed, set to position 2, and kept closed for a long enough time so that the capacitors become fully charged, they can be taken out of the circuit to study the resistive aspects of the circuit (Figure 24.13).

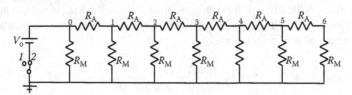

FIGURE 24.13 The resistive circuit for the study of the electrical properties of the axon.

If the voltage difference between the ground and each of the numbered (0–6) points on the circuit is measured, the effect of resistive elements of the circuit on the signal as it travels down the axon can be studied.

The first segment of the resistive circuit, shown in Figure 24.14, has the voltage of the battery, V_o, across point 0 and ground, which is dropped across only one resistor, R_M. Employing Kirchhoff's Voltage Law: $V_o - I_0 R_M = 0$, so the voltage at point 0 is just the voltage of the battery, V_o, so $V_0 = V_o$

FIGURE 24.14 The first segment of the resistive circuit.

For the measurement of the voltage at point 1, the current I_1 in Figure 24.15 must be found.

$$\text{Loop A: } V_o - I_0 R_M = 0$$
$$\text{Loop B: } V_o - I_1 R_A - I_1 R_M = 0$$
$$\text{Node 0: } I = I_0 + I_1$$

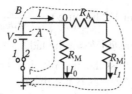

FIGURE 24.15 The second segment of the resistive circuit.

From Loop B: $I_1 = V_o/(R_A + R_M)$

From Ohm's Law: $V_1 = I_1 R_M$, so substituting the expression of I_1 from Loop B's results into Ohm's Law gives:

$V_1 = [V_o/(R_A + R_M)] R_M$ or for an expression of V_1 as a function of V_o,

$$V_1 = V_o\left(\frac{R_M}{(R_A + R_M)}\right) = V_o R_m\left(\frac{1}{(R_A + R_M)}\right)$$

The next node in Figure 24.13 is number 2, so the next sub-circuit to analyze is given in Figure 24.16. The two loops are labeled A and B in Figure 24.16, but numbers have been added to the subscripts to explain the loops, and then they will be dropped because all R_As are equal and all R_Ms are equal.

Loop A: $V_o - (I_1 + I_2)R_{A1} + I_1 R_{M1} = 0$

Loop B: $V_o - (I_1 + I_2)R_{A1} + I_1(R_{A2} + R_{M2}) = 0$

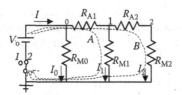

FIGURE 24.16 The third segment of the resistive circuit.

Notice, that the combination of I_1 and I_2 flows through R_{A1}. Dropping all the subscript numbers for the resistors and solving both loops for I^1 gives:

Loop A: $I_1 = [V_o - (I_2 R_A)]/[(R_A + R_M)]$

Loop B: $I_1 = [V_o - I_2(2R_A + R_M)]/(R_A)$

Set the right-hand side of each loop equation equal to the other since they both equal I_1, which results in:

$$[V_o - (I_2 R_A)]/[(R_A + R_M)] = [V_o - I_2(2R_A + R_M)]/(R_A)$$

This expression can be solved for I_2 to get: $I_2 = (V_o R_M)[1/[(R_A + R_M)(2R_A + R_M) - R_A^2]$

Multiplying out the quantities in the expression above and applying Ohm's Law, the voltage measured at point 2 in the circuit in Figure 24.16 is

$$V_2 = I_2 R_M = V_o R_M\left(\frac{R_M}{R_A^2 + 3R_A R_M + R_M^2}\right),$$

Notice that V_2 is a smaller value than V_1, because $\left(\frac{R_M}{R_A^2 + 3R_A R_M + R_M^2}\right) < \left(\frac{1}{(R_A + R_M)}\right)$ and V_1 is smaller than V_o, because $\left(\frac{1}{(R_A + R_M)}\right) < 1$. Each step down the ladder circuit, because it looks like a ladder, results in a reduction of the voltage across the next resistor. The reduction of the voltage across the next numbered node down the ladder circuit is not linear.

If measurements of voltage are made along the positions numbered 0 to 6 along the ladder of the circuit in Figure 24.13 and plotted, a graph like the one in Figure 24.17 will result.

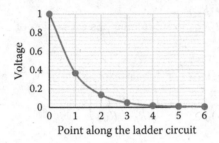

FIGURE 24.17   Graph of voltage vs points along the ladder circuit.

With x as the position along the axon corresponding to node 0 to 6 in the circuit in Figure 24.13, the voltage as a function of x can be expressed as equation (24.7) as

$$V_n = V_o e^{-\frac{x}{\lambda}}. \tag{24.7}$$

where, in neurobiology, λ is known as *the length constant*. In SI units, λ has units of meters. When the signal has traveled a distance $x = \lambda$ down the ladder, the voltage is:

$$V = V_o e^{-\frac{\lambda}{\lambda}} = (e^{-1})V_o \cong .37 \; V_o$$

24.5.4   TEMPORAL RESPONSE OF THE LADDER CIRCUIT

A signal is produced by rapidly closing and opening the switch in the ladder circuit in Figure 24.11. The term *closing* refers to a switch that will conduct electricity, so it is in position 2 in Figure 24.11. The term *opening* refers to a switch that cannot conduct electricity, for example, in position 1 in Figure 24.11. If the switch in Figure 24.11 is initially open, at position 1, the voltage across the capacitor will be zero. If at the start of the signal, at $t = 0$ s, the switch is quickly closed, moved to position 2 in Figure 24.11, the voltage will quickly rise to Vo. If the switch is left closed, position 2, for 4 ms and then suddenly opened, switched to position 1, a signal similar to a square wave, demonstrated in the Figure 24.18, will be produced.

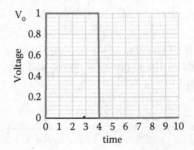

FIGURE 24.18   Square wave input voltage.

If a square wave signal is generated by opening and closing the switch described in this paragraph and the voltage across point 0 and ground is monitored, the signal similar to the one in Figure 24.19 will be measured.

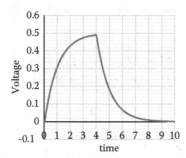

FIGURE 24.19 RC signal across the ladder circuit.

Notice that the signal increases like a charging RC circuit and decreases like a discharging RC circuit.

Even the simplest node to analyze, the capacitor at the 0 node, has several pathways from which to discharge. To get a sense of the analysis, this example will use only the first two pathways that the current can take from the capacitor to ground for discharging.

Remember, for a discharging capacitor in an RC circuit, the time dependence of the voltage can be represented as equation (24.5)

$$V = V_o \left(e^{-\frac{t}{\tau}} \right)$$

with $\tau = RC$. The C is just the capacitance of the capacitor, which was determined to be $C_M = 5.4062 \times 10^{-11}$ F for this system. The R is a combination of the resistance of the possible discharge paths. The two primary pathways for charge to flow to ground from the positive plate of the capacitor at node 0 are shown Figure 24.20.

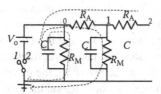

FIGURE 24.20 Discharging of the 0-node capacitor.

The first pathway is through the R_M in parallel with the capacitor, and the next is up through R_A and then down through R_M to ground. The current cannot flow through the battery because the switch is open, so that pathway is open. This is the part of the square wave in which there is no voltage from the source. Notice that these two pathways are parallel to each other, whereas R_A and R_M are in series relative to the top pathway that passes through points 0 and 1 and R_A at the top and down through R_M. The resistance of these two pathways can be found by first adding R_A and R_M over the top path,

$$R_{AM} = R_A + R_M$$

and then adding these two parallel pathways reciprocally.

$$(1/R_{eq}) = (1/R_{AM}) + (1/R_M)$$

This R_{eq} is the R in the calculation of the time constant, and given the values of R_M and R_A as $R_A = 1.061 \times 10^6 \ \Omega$, $R_M = 1.1459 \times 10^7 \ \Omega$, the value of this R_{eq} can be found in the following steps.

Find: $R_{AM} = R_A + R_M = 1.061 \times 10^6 \ \Omega + 1.1459 \times 10^7 \ \Omega = 1.252 \times 10^7 \ \Omega$

Find: $(1/R_{eq}) = (1/R_{AM}) + (1/R_M) = (1/1.252 \times 10^7 \ \Omega) + (1/1.1459 \times 10^7 \ \Omega)$, gives

$R_{eq} = 5.983 \times 10^6 \ \Omega$

This results in a time constant of:

$$\tau = R_{eq}C = (5.983 \times 10^6 \ \Omega)(5.4062 \times 10^{-11} \ \text{F}) = 3.235 \times 10^{-4} \ \text{s}$$

This is close to the time of 0.5 ms (0.5×10^{-3} s = 5×10^{-4} s) it takes a signal to travel down a synapse. A more detailed study of the complete RC ladder circuit analysis requires some complex math, so this analysis ends here, but the hope is that it will inspire you to further study this interesting application of circuit theory in biology. Much of the information for this section was taken from the bibliography at the end of this chapter.

24.6 QUESTIONS AND PROBLEMS

24.6.1 MULTIPLE CHOICE QUESTIONS

1. In Figure 24.2, the battery has a voltage if V_o, the capacitor has a capacitance C and is uncharged, the resistance of the resistor is R, and the switch is initially at position 2 for a long enough time to discharge the capacitor. At time $t = 0$ s, the switch is moved from position 2 to position 1. At the first instant after the switch is moved from position 2 to position 1, what happens?
   A. The charge on the capacitor is 0, and the current through the resistor is V_o/R.
   B. The charge on the capacitor is $C V_o$, and the current through the resistor is 0.
   C. The charge on the capacitor is $C V_o$, and the current through the resistor is V_o/R.
   D. The charge on the capacitor is 0, and the current through the resistor is 0.
2. In Figure 24.2, the battery has a voltage if V_o, the capacitor has a capacitance C and is uncharged, the resistance of the resistor is R, and the switch is initially at position 2 for a long enough time to discharge the capacitor. At time $t = 0$ s, the switch is moved from position 2 to position 1. At the first instant after the switch is moved from position 2 to position 1, what happens?
   A. The current through the resistor and the voltage across the capacitor both decrease.
   B. The current through the resistor decreases, and the voltage across the capacitor increases.
   C. The current through the resistor increases, and the voltage across the capacitor decreases.
   D. The current through the resistor and the voltage across the capacitor both increase.
3. In Figure 24.2, the battery has a voltage if V_o, the capacitor has a capacitance C and is uncharged, the resistance of the resistor is R, and the switch is initially at position 2 for a long enough time to discharge the capacitor. At time $t = 0$ s, the switch is moved from

position 2 to position 1. At the first instant after the switch is moved from position 2 to position 1, what happens?

A. The voltage across the resistor and the voltage across the capacitor both start out at 0 and increase, asymptotically approaching V_o.

B. The voltage across the resistor and the voltage across the capacitor both start out at V_o and decrease, asymptotically approaching 0.

C. The voltage across the resistor starts out at 0 and increases, asymptotically approaching V_o. The voltage across the capacitor starts out at V_b and decreases, asymptotically approaching 0.

D. The voltage across the capacitor starts out at 0 and increases, asymptotically approaching V_o. The voltage across the resistor starts out at V_o and decreases, asymptotically approaching 0.

4. In Figure 24.6, the 10 µF capacitor is fully charged by a 6 V battery, because the switch is in position 1 for a long enough time. At $t = 0$ s, the fully charged capacitor is discharged through resistor 2 to ground. This process is repeated for three resistors with resistances of 100 kΩ, 500 kΩ, and 1 MΩ, respectively. Through which resistor does it take the greatest amount of time for the capacitor to discharge?

A. 100 kΩ

B. 500 kΩ

C. 1 MΩ

D. All take the same amount of time.

5. In Figure 24.6, the capacitor is fully charged by a 6 V battery, because the switch is kept in position 1 for a long enough time. At $t = 0$ s, the fully charged capacitor is discharged through resistor 2 to ground. This process is repeated for 3 different versions of the circuit, all with $R_2 = 1$ MΩ and 3 different capacitors with values of: 2.2 µF, 4.7 µF, and 10 µF. Which capacitor takes the most time to discharge?

A. 2.2 µF

B. 4.7 µF

C. 10 µF

D. All take the same amount of time.

6. In Figure 24.6, the capacitor is fully charged by a 12 V battery, R_2 is initially 200 kΩ, and C is initially 20 µF. The switch is set to position 1 for a long enough time so that the capacitor is fully charged. The switch is moved to position 2 at $t = 0$ s, and the time for the voltage across the capacitor to decrease from 12 V to 4.44 V is measured to be 4 s. If the resistance of R_2 in the circuit is increased to 300 kΩ, would the time it takes the capacitor to discharge from its maximum voltage to 4.44 V be greater than, less than, or equal to 4 s?

A. greater than

B. less than

C. equal to

7. In Figure 24.6, the capacitor is fully charged by a 12 V battery, R_2 is initially 200 kΩ, and C is initially 20 µF. The switch is set to position 1 for a long enough time so that the capacitor is fully charged. The switch is moved to position 2 at $t = 0$ s, and the time for the voltage across the capacitor to decrease from 12 V to 4.44 V is measured to be 4 s. If the capacitance of C in the circuit is decreased from 20 µF to 10 µF, will the maximum charge that the capacitor stores be greater than, less than, or equal to the maximum charge on the 20 µF in the original circuit?

A. greater than

B. less than

C. equal to

8. Assuming the resistivity of the axoplasm to be the same in both cases, how does the axoplasm resistance R_{A1} of a 1 mm length of an axon of inner diameter 6 μm compare with the axoplasm resistance R_{A2} of a 1 mm length of an axon of inner diameter 3 μm?
   A. $R_{A1} = \frac{1}{4}R_{A2}$
   B. $R_{A1} = \frac{1}{2}R_{A2}$
   C. $R_{A1} = R_{A2}$
   D. $R_{A1} = 2R_{A2}$
   E. $R_{A1} = 4R_{A2}$

9. Assuming the resistivity and thickness of the membrane to be the same in both cases, how does the membrane resistance R_{M1} of a 1 mm length of an unmyelinated axon of inner diameter 6 μm compare with the membrane resistance R_{M2} of a 1 mm length of an unmyelinated axon of inner diameter 3 μm?
   A. $R_{M1} = \frac{1}{4}R_{M2}$
   B. $R_{M1} = \frac{1}{2}R_{M2}$
   C. $R_{M1} = R_{M2}$
   D. $R_{M1} = 2R_{M2}$
   E. $R_{M1} = 4R_{M2}$

10. Assuming the dielectric constant and thickness of the membrane to be the same in both cases, how does the membrane capacitance C_{A1} of a 1 mm length of an unmyelinated axon of inner diameter 6 μm compare with the membrane capacitance C_{A2} of a 1 mm length of an axon of inner diameter 3 μm?
   A. $C_{A1} = \frac{1}{4}C_{A2}$
   B. $C_{A1} = \frac{1}{2}C_{A2}$
   C. $C_{A1} = C_{A2}$
   D. $C_{A1} = 2C_{A2}$
   E. $C_{A1} = 4C_{A2}$

24.6.2 PROBLEMS

1. Referring to Figure 24.2, assume a battery voltage of $V_o = 17.0$ V, $R_1 = 1.50$ MΩ, and $C = 1.80$ μF. The switch is set to position 2 for a long enough time so that the capacitor is completely discharged. At $t = 0$ s, the switch is moved to position 1, and the capacitor begins to charge.
   a. Calculate the time constant of the RC circuit.
   b. Find the maximum charge that will be stored on the capacitor during charging.
   c. How long after the switch is moved from position 2 to position 1 does it take for 10.0 μC of charge to build up on the capacitor? *Hint: to do this problem you will need to solve the equation for t and remember that ln(x) is the inverse of e^x.*

2. Referring to Figure 24.2, assume a battery voltage of $V_o = 200.0$ V, $R_1 = 100$ Ω, and $C = 20$ mF. The switch is set to position 2 for a long enough time so that the capacitor is completely discharged. At $t = 0$ s, the switch is moved to position 1, and the capacitor begins to charge.
   a. What is the time constant of the circuit?
   b. Find the voltage across the capacitor at $t = 1$ s.

3. Referring to Figure 24.2, assume a battery voltage of $V_o = 100.0$ V, $R_1 = 10$ Ω, and $C = 2$ mF. The switch is set to position 2 for a long enough time so that the capacitor is completely discharged. At $t = 0$ s, the switch is moved to position 1, and the capacitor begins to charge.
   a. What is the time constant of the circuit?
   b. Find the voltage across the capacitor at $t = 0.1$ s.

4. Referring to Figure 24.2, assume a battery voltage of $V_o = 200.0$ V, $R_1 = 100$ Ω, and $C = 20$ mF. The switch is set to position 2 for a long enough time so that the capacitor is completely discharged. At $t = 0$ s, the switch is moved to position 1, and the capacitor begins to charge. The switch is left at position 1 for a long enough time so that the capacitor is fully charged. If the switch is then switched back to position 2 so that the capacitor discharges, how long after the switch is put back to position 2 does the capacitor have half its maximum charge?

5. Referring to Figure 24.6, assume a battery voltage of 6.15 V, $R_2 = 1$ MΩ, and $C = 8.85$ μF. The switch is set to position 1 for a long enough time so that the capacitor is fully charged. If the switch is moved to position 2 at $t = 0$ s, how long does it take the voltage to reach .5% of its maximum value?

6. Referring to Figure 24.6, assume a battery voltage of 12 V, $R_2 = 22.5$ MΩ, and $C = 47$ nF. The switch is set to position 1 for a long enough time so that the capacitor is fully charged. The switch is moved to position 2 at $t = 0$ s. Find the charge on the capacitor at time $t = 1.25$ s.

7. Referring to Figure 24.6, assume a battery voltage of 12 V, $R_2 = 200$ kΩ, and $C = 20$ μF. The switch is set to position 1 for a long enough time so that the capacitor is fully charged. The switch is moved to position 2 at $t = 0$ s. Find the charge on the capacitor at time $t = 10$ s.

8. Referring to Figure 24.6, assume a battery voltage of 10 V, $R_2 = 4$ kΩ, and $C = 5$ μF. The switch is set to position 1 for a long enough time so that the capacitor is fully charged. The switch is moved to position 2 at $t = 0$ s. Find the charge on the capacitor at time $t = 8$ ms.

9. Referring to Figure 24.6, assume a battery voltage of 10 V, $R_2 = 2$ kΩ, and $C = 3$ μF. The switch is set to position 1 for a long enough time so that the capacitor is fully charged. The switch is moved to position 2 at $t = 0$ s. Find the charge on the capacitor at time $t = 15$ ms.

10. It is generally said that the specific capacitance c, that is, the capacitance per area of cell membrane, is $1\frac{\mu F}{(cm)^2}$ (given only to one significant figure, so we should consider it an approximate value). Using the axon membrane dielectric constant of $\kappa = 7$ as well as the axon membrane thickness ($b = 6$ nm), calculate the specific capacitance c in units of $\frac{\mu F}{(cm)^2}$ for comparison with the generally accepted value of $1\frac{\mu F}{(cm)^2}$.

BIBLIOGRAPHY

1. Herman, Irving P., *Physics of the Human Body*, Springer International Publishing, Switzerland, 2006.
2. Hobbie, Russell K. and Roth, Bradley J., *Intermediate Physics for Medicine and Biology*, 5th ed., Springer International Publishing, Switzerland, 2015.
3. Robinson, M., Martin, J., Atwood, H., and Cooper, R., *Modeling Biological Membranes with Circuit Boards and Measuring Electrical Signals in Axons: Student Laboratory Exercises*. National Library of Medicine, Published online 2011 Jan 18. doi: 10.3791/2325.
4. Susuki, K., "Myelin: A Specialized Membrane for Cell Communication", *Nature Education Communications*, 3(9): 59, 2010.

25 Light and Color

25.1 INTRODUCTION

This is the introductory chapter on the study of light, which is known as optics. The chapter begins with an introduction to the photon and the wave nature of light. The energy associated with these representations of light are discussed, and examples are provided. The perception of color is discussed, along with examples of color addition and color subtraction. Next, the basics of ray optics are introduced through the laws of reflection and refraction. Examples of the bending of light in prescribed ways are provided in the context of a prism. This discussion provides an opportunity to discuss Newton's experiments on light and color.

25.2 CHAPTER QUESTION

You may already know that when sunlight is shown through a prism a rainbow of colors, called the visible spectrum, results. The colors of light, red, orange, yellow, green, blue, indigo, and violet, always appear in the same order of the visible spectrum, as shown in Figures 25.1.

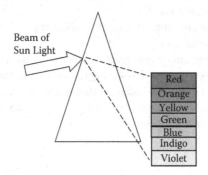

FIGURE 25.1 Colors of light from a prism.

The chapter question is what property of light determines the colors we see and how does a prism "create" this spectrum? This question will be answered at the end of the chapter employing the concepts developed throughout the chapter.

25.3 LIGHT WAVES AND PHOTONS

To answer the question about color and the property of light we associate with color, the basic theories of light must be understood. In some situations, the characteristics of light can best be explained if light is considered to be a wave of electric and magnetic fields and in other situations a particle of energy model of light is best. For example, if a yellow light is generated by a sodium lamp, the particle description of light provides the best description of how the light is produced. If we are studying the colors seen when sunlight scatters off a diffraction grating, like when you see all the colors scattered off a CD or DVD, it is best to use a wave model of light. This dual explanation of light is called the ***wave-particle duality***. In this chapter, each theory will be used at a basic level, and in later chapters each model of light will be investigated in more depth when applied to understand important physical phenomena.

DOI: 10.1201/9781003308072-26

25.3.1 Sources of Light

The source of all light is accelerating charged particles, which in most cases are electrons. In solids and liquids, the charges vibrate and thus generate light. It is common to switch from (f) to the Greek letter "nu" (ν) as the variable for the frequency of a light wave. This is done to make a distinction between the frequency (ν) of the light wave and the frequency (f) of the vibrating charges that generated the light wave. As depicted in Figure 25.2, the charged particles in solids and liquids can be thought of as attached together by springs, which represent the bonds between the atoms.

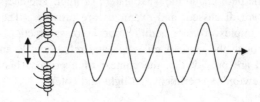

FIGURE 25.2 A vibrating charged particle in a solid generating a light wave.

As solids and liquids are heated, the atoms and molecules vibrate at a higher frequency (f), so the light waves produced have a higher frequency (ν). From an energy point of view, some of the additional thermal energy that causes an object to be hotter also results in faster vibrations of the atoms and thus higher energy photons.

As a piece of metal is heated up, it begins to glow red, then yellow, and maybe even white hot as the particles vibrate faster and faster.

The explanation given in the above paragraph is known as the black body theory. It explains that the color of light that we see is related to the energy of the photons and/or the frequency of the light waves. Wein's Law is an empirical equation that relates the center wavelength (λ_{max}) of the curve to the temperature of the object radiating the light. Figure 25.3 shows the black body curve of two objects heated to different temperatures.

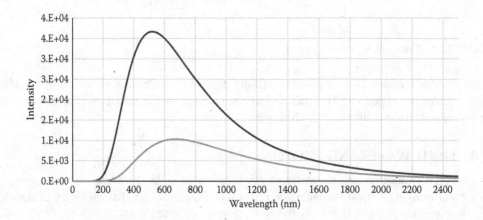

FIGURE 25.3 Black body curves at 5800 K and 4500 K in arbitrary units of intensity.

In this plot, the 5800 K curve has a greater intensity, and it is centered at a wavelength of approximately 530 nm. The curve with a smaller intensity is for an object with a surface temperature of 4500 K, and it has a center wavelength of 670 nm. The surface of the sun is at a temperature of 5800 K, so the color of the sun is centered at almost 530 nm. The filament in an incandescent light bulb is about 3000 K, so it is centered at 970 nm. Most of its energy is in the infrared (heat). This is why these bulbs are so hot and inefficient and also why they glow more orange-red than sunlight. Notice that the greater the temperature, the farther to the left, or the shorter wavelength, the peak wavelength curve is located.

Wein's Law, expressed in equation (25.1), is an empirical equation that relates the center wavelength (λ_{max}) of the black body curves to the temperature of the object radiating the light that produced the curve.

$$\lambda_{max} = 2.8977685 \times 10^3 \, \frac{mK}{T\,(K)} \tag{25.1}$$

The units of the constant in the numerator of the equation are mK (meters times the temperature of the object in Kelvin), so to use Wein's Law the temperature of the object must be measured in the Kelvin scale.

25.3.2 WAVE THEORY OF LIGHT

Recalling back to Chapter 15, waves are characterized with the quantities of amplitude (A), frequency (f), wavelength (λ), and period (T). Also, through the definitions of these quantities a simple relationship between the frequency, wavelength, and velocity (v) of a wave was generated, which is repeated here as equation (25.2):

$$v = \lambda f \tag{25.2}$$

Through many experiments and theoretical calculations, the speed of light in a vacuum has been established with a value of 299,792,458 m/s, which for all calculations done in this text, the value will be labeled with a c and rounded to $c = 3 \times 10^8$ m/s.

The speed of light in air is a bit slower than the speed of light in a vacuum. It is 299,702,547 m/s, which is also rounded to 3×10^8 m/s for the calculations for light traveling through air in this text. In addition, it is common to use the Greek letter (v) as the variable for the frequency of a light wave, so the equation that relates the speed of light in a vacuum or air, the wavelength, and the frequency of the light is given in equation (25.3) as

$$c = \lambda v \tag{25.3}$$

Example

Find the wavelength of a light wave traveling through air, which has a frequency of 5×10^{14} Hz.
Starting with equation (25.3) $c = \lambda v$ and solving it for wavelength gives:

$$\lambda = c/v = (3 \times 10^8 \text{ m/s})/(5 \times 10^{14} \text{ m})$$
$$\lambda = 6 \times 10^{-7} \text{ m} = 600 \times 10^{-9} \text{ m} = 600 \text{ nm}$$

This is the wavelength of light that looks orange-yellow.

25.3.3 PARTICLE THEORY OF LIGHT

In 1905, Albert Einstein published a paper titled "Concerning an Heuristic Point of View Toward the Emission and Transformation of Light" in which he introduced the concept that light may be made up of individual, quantized bundles of energy. In physics, *quantized* means the quantity that is quantized can only come in specific amounts. Today these quantized bundles of light energy are called photons. In this paper, Einstein developed a simple relationship between the energy, E, of the photons of light and the frequency, v, of the light waves, which is given in equation (25.4) as

$$E = hv. \tag{25.4}$$

Since Einstein built upon the work of Max Planck, the constant of equality between these properties of light is called Planck's constant: $h = 6.626 \times 10^{-34}$ J s.

Example

Find the energy of a light wave traveling through air, which has a frequency of 5×10^{14} Hz nm.

$$E = hv = (6.626 \times 10^{-34} \text{ J s})(5 \times 10^{14} \text{ Hz}) = 3.313 \times 10^{-19} \text{ J}$$

This is the energy of a photon in a beam of light that has a wavelength of 600 nm and looks orange.

In 1917, Einstein published a second paper on the topic, titled "On the Quantum Theory of Radiation". In this paper, Einstein formalized the concept of the photon and described interactions between atoms and photons with his new theory. The ideas from this paper have led to the development of technologies such as lasers and x-rays. Although Einstein may be best known for his theories of relativity, this concept of the photon won Einstein the 1921 Nobel Prize in Physics.

25.4   COLOR

Employing the two theories, today it is understood that the color of light that we see is related to the energy of the photons, which is associated with the frequency of the light waves. Since the wavelength and the frequency of a light wave are linked together by the speed of the light wave, and the wavelengths of the light in nm are easier to remember, the wavelengths in nm and photon energies in $(\times 10^{-19})$ Joules are given in Table 25.1.

TABLE 25.1
Color, Wavelength, and Photon Energy

Color	Wavelength (nm)	Photon Energy ($\times 10^{-19}$ J)
Red	650	3.058
Orange	600	3.313
Yellow	580	3.427
Green	530	3.751
Blue	470	4.229
Indigo	440	4.518
Violet	400	4.970

All the colors of light are present in the sunlight that illuminates our planet. When this light is shown upon a piece of plain paper, we call it white. Thus, we refer to light that contains all the colors as white light. Conversely, the absence of all colors of light is called black.

In the same way that color can be considered in both the photon and the wave model of light, the brightness of light is related to both the number of photons in the beam of light and the *amplitude* of the light waves. Brightness is measured as intensity (I), which is defined as power per area, as expressed in equation (25.5) as

$$I = \frac{P}{A}. \qquad (25.5)$$

Therefore, the units of intensity are W/m^2. In either the photon or the wave model of light, there are connections between the number of photons and the amplitude of the E&M wave and the intensity.

25.4.1 THE HUMAN EYE AND COLOR

The human eye contains two types of cells that are sensitive to light. The rod cells are sensitive to any color of visible light and thus give us the ability to see objects illuminated by most visible wavelengths. The rods do not play a large part in the discrimination of objects based on color. The sensitivity of the rod cells is centered in the blue-green and extends across the entire visible spectrum. They may not be helpful in distinguishing color, but they are very sensitive, so they are useful in dim light. That is why no matter what color an object is when we observe it in very dim light it may appear dull gray.

In contrast to the rod cells, the three different kinds of cone cells have sensitivities centered on three different wavelengths, respectively, roughly corresponding to the colors blue, green, and red, with sensitivity curves labeled S, M, and L, corresponding to short, medium, and long wavelengths, respectively. Thus, when our retina is illuminated with blue light with a wavelength of 450 nm, the S (blue) cone is stimulated to 100%, and the M (green) and L (red) cones are barely stimulated. The signal is sent to our brain, and we were taught this is called blue. It is important to recognize that the sensitivities of the three different types of cones overlap. Thus, when we see yellow light with a wavelength of 580 nm, both the M (green) and L (red) cones are stimulated to approximately 95%. This signal is sent to our brain, and we were taught that this is yellow.

25.4.2 COLOR ADDITION

It is not a coincidence that the three colors of pixels in your TV screen and computer monitors are red ($\lambda \sim 660$ nm), green ($\lambda \sim 532$ nm), and blue ($\lambda \sim 440$ nm). The TV utilizes the RGB color model of additive mixing of colors. There are no yellow pixels on the TV, so to produce an image of a yellow object on the screen, the red and green pixels are both activated. Thus, both your M and L cones are stimulated. We cannot tell the difference between the pure yellow light with a wavelength of 580 nm from a sodium lamp, and the yellow of an image of a sodium lamp on the TV (an image that contains no light at 580 nm, only equal amounts of light at 532 nm and 660 nm). So, what does yellow mean? Sometimes it is in your brain, and other times it is a property of the light itself. We cannot tell just by looking. We need a device like a prism to sort this out.

25.4.3 COLOR SUBTRACTION

Objects that do not produce light, but can be seen, must reflect light. The reflection of light off an object can be thought of as either light waves bouncing off objects like the echo of a sound wave or photons bouncing off atoms. This "bouncing" is not exactly the case, but it gives us a conceptual framework from which to start, and we will readdress this in a later chapter. With either conceptual model, the color we see an object to have is based on the colors of light the object reflects. Thus, plants appear green because they reflect green light and absorb blue and red light. Note that plants reflect light having the most plentiful wavelengths of sunlight.

The colors of objects that do not generate light result from the colors of light that such objects absorb, and thus subtract, from white light. The primary colors of subtraction are denoted as the colors that result when one of the additive primary colors of light is subtracted from white light.

If red is subtracted from white light, blue and green light remain, and the mixture of these colors is *cyan*. If blue is subtracted from white light, red and green light remain, and the mixture of these colors is *yellow*. If green is subtracted from white light, blue and red light remain, and the mixture of these colors is *magenta*. The next time you change the ink in your color printer, notice that these are the colors of ink used to generate all the colors in our colored printouts.

25.4.4 THE COLOR WHEEL

A pneumonic device that can be used to help keep these color combinations straight is the color wheel, which is depicted in Figure 25.4.

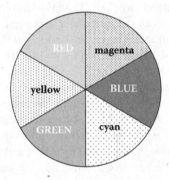

FIGURE 25.4 The color wheel.

The color wheel is a circle divided into six equal-sized segments. To construct the color wheel, the three primary colors of light, in capital letters and solid color, are put into alternating segments of the circle so that the segments are labeled, starting with RED, space, BLUE, space, GREEN, and space around the circle. In the space between RED and BLUE, magenta is written since (RED + BLUE) light = magenta light, and the piece of the circle is filled in with a texture to represent a complimentary color of subtraction. In the space between BLUE and GREEN, cyan is written since (BLUE + GREEN) light = cyan, and the piece of the circle is filled in with a texture to represent a complimentary color of subtraction. In the space between RED and GREEN, yellow is written since (RED + GREEN) light = yellow, and the piece of the circle is filled in with a texture to represent a complimentary color of subtraction. The color wheel is helpful in answering conceptual questions about light. If you are drawing this yourself, you can use colored pencils and/or crayons to produce it in color.

25.4.5 COLOR WHEEL EXAMPLES

Example 1

A large body of water appears cyan, sometimes referred to as aqua blue in this context. This color is most obvious in a large swimming pool with white bottoms and sides. It is also seen in the color of water in the Caribbean. What color does the water absorb?

Answer: Referring to the color wheel in Figure 25.5, since water absorbs red light the "Not Allowed" symbol is put into the RED section of the color wheel.

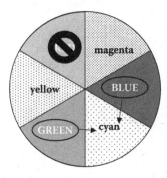

FIGURE 25.5 The color wheel for example 1.

This means that when all the colors of the white light from the sun strike the water, the RED light is absorbed and BLUE and GREEN light are reflected. A combination of BLUE and GREEN triggers a signal that our brain tells us is the color cyan.

Example 2

When cyan and yellow ink is mixed as a spot on a page, what color is observed?

Since cyan ink absorbs the red light a "Not Allowed" symbol is put into the RED section of the color wheel in Figure 25.6, and since yellow ink absorbs blue light a "Not Allowed" symbol is put into the BLUE section of the color wheel in the same figure.

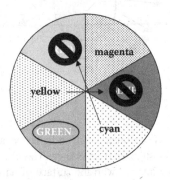

FIGURE 25.6 The color wheel for example 2.

Therefore, the spot on the paper will appear GREEN. Another way to see this is that GREEN is between cyan and yellow on the color wheel. Either way, the incoming white light is a mixture of red, green, and blue light. Cyan ink absorbs the RED light, and yellow ink absorbs the BLUE light, so the only color of light reflected is GREEN.

Example 3

Color Addition: If there is an image of a ripe banana on your computer screen, what pixels are emitting light from the computer screen to produce the image of the yellow banana? Note that the computer screen is a light emitter with only red, green, and blue pixels. So, with the RED and GREEN pixels illuminated, signified by the ovals around the colors with the arrows pointing to yellow in Figure 25.7, our brain will interpret this as yellow.

The color wheel is a helpful pneumonic device to investigate the combination of colors either by addition or subtraction.

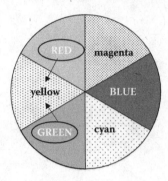

FIGURE 25.7   The color wheel for example 3.

25.5   REFLECTION AND REFRACTION

Another fundamental property of light is that it travels in straight lines. These straight paths of light are indicated by vectors, called rays, coming from a light source. For example, the light bulb depicted in Figure 25.8 has an infinite number of rays emanating from the bulb in all directions.

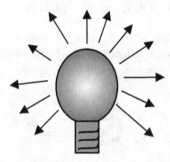

FIGURE 25.8   Light rays from a light source.

Remember that the light bulb is three dimensional, so light rays also come out of and into the page. Picture a great number of toothpicks stuck into an orange. These light rays will travel in straight lines through a vacuum forever until they interact with the surface of an object or an interface with another material. When light rays interact at the interface between two materials, some of the light is reflected off the interface between the surfaces, and some is transmitted into the material, as depicted in Figure 25.9.

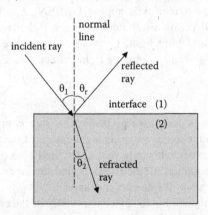

FIGURE 25.9   Reflection and refraction of light at an interface between two materials.

It is common to label the two materials as (1) and (2), as shown in Figure 25.9. The ray moving toward the interface between the material is called the incident ray, and the line that starts at the point at which the incident ray strikes the surface of the object and is perpendicular to the surface is called the normal line. Remember that the Normal Force received its name because it is perpendicular to the surface. This normal line gets its name from the mathematical word meaning perpendicular. The angle between the incident ray and the normal line is labeled θ_1.

25.5.1 REFLECTION

The reflected ray is at an angle of θ_r, as measured from the normal line, and the transmitted ray is called the refracted ray, which is at an angle, θ_2, as measured from the normal line on the other side of the interface.

Referring to Figure 25.9, it is a fact of nature, called the **law of reflection,** that the angle of incidence, θ_1, is equal to the angle of reflection, θ_r, and is expressed in equation (25.6) as

$$\theta_1 = \theta_r. \tag{25.6}$$

25.5.1.1 Types of Reflections of Light

The law of reflection is true for all types of surfaces, whether smooth or rough. So why is it that smooth and rough surfaces look different to us? The explanation lies not in the reflection of each light ray striking the surface, but the collective result of a large number of rays, each following the law of reflection. To describe the two kinds of reflection, we start with a group of light rays, all parallel to each other and traveling toward a surface. As each of the incident rays from the light strikes a smooth surface, like the glass of mirror, they reflect at the same angle. This type of reflection, in which each incident ray of light bounces off in a single direction, is called *specular reflection,* which is demonstrated on the left side of Figure 25.10. If you look at the surface, you will not only see the light from the bulb, but the image of the light bulb itself. So, with specular reflection, you can see the image of the light source.

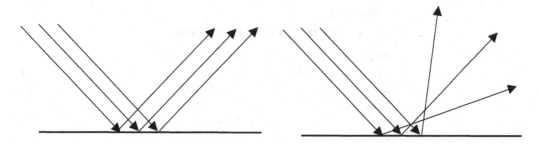

FIGURE 25.10   Specular and diffuse reflections.

If the same group of parallel light rays is incident on a rough object, like the surface of the road, the reflected rays will move off at different angles. The law of reflection still holds for each ray, but since the surface is rough, the normal line for each ray will be in a slightly different direction. So, the angles of incidence and reflection will also be different for each ray. The result when you look at the road is that you see the road, because light is reflecting off of it. You will not see an image of the light source because the light rays are scattered in different directions and many do not reach your eyes. This type of reflection, in which the rays move off at different angles, is known as a *diffuse reflection,* which is demonstrated on the right side of Figure 25.10.

So, as objects become smoother, you begin to see the object reflected from their surfaces more clearly. If you polish a smooth table surface, sometimes you can actually see a reflection of the light bulb in the lamp. Since most surfaces are rough, at the microscopic level, most of what we see in the world is a result of diffuse reflection.

25.5.2 REFRACTION

As demonstrated in Figure 25.8, the portion of the light ray that passes from one material into another bends relative to the original path of the incident light ray. If it is assumed that the light ray is starting in air and entering into a block of glass, the light ray bends closer to the normal line. The relationship between the incident angle (θ_1) and the refracted angle (θ_2) at the interface was first empirically, through experiment, discovered by Willebrord Snell in 1621. Snell measured the angles, θ_1 and θ_2, for different materials, and he developed an equation to represent the relationship of the angles that is called Snell's Law and is given by equation (25.7) as

$$n_1 \sin(\theta_1) = n_2 \sin(\theta_2) \tag{25.7}$$

where n_1 and n_2 are known as the index of refraction for a specific material. This expression does not try to explain the bending of the light at the interface, only that there is bending and the angles follow this expression. The value of the index of refraction of a material represents the degree to which a light ray bends as it travels from a vacuum into a material. Thus, the index of refraction of a vacuum is set at a value of 1.0, and the bending of light in other materials is based off this quantity. Light moving from a vacuum into any other material will bend so that the angle of refraction will be less than the angle of incidence. So, the index of refraction of any material is greater than 1, and the greater the index of refraction, the more light will bend when it enters this material. The indices of refraction for some common materials are given in Table 25.2.

TABLE 25.2
Index of Refraction of Some Materials

Material	Index of Refraction
Vacuum	1.0
Air	1.0001
Water	1.33
Ice	1.31
Glass	1.5
Diamond	2.42

Since light traveling from the vacuum of space into the earth's atmosphere does not bend much, the index of refraction of air is approximately one (1.0001). On the other hand, light rays are bent a noticeable amount by water, which has an index of refraction of 1.33. Diamonds, which have an index of refraction of 2.42, bend light so strongly that the light rays get trapped in the crystal if cut correctly and the diamond appears to sparkle. Index of refraction is one of the best ways to make sure the diamond is real.

Example

If a ray of light traveling through air is incident on a smooth surface of glass at an angle of 30° to the normal, as shown in Figure 25.11, what will be the angle of refraction of the light ray that enters the glass?

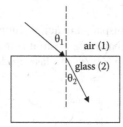

FIGURE 25.11 Refraction example.

Solution

$$n_1 \sin(\theta_1) = n_2 \sin(\theta_2)$$

$$(1.0001) \sin(30°) = (1.5)\sin(\theta_2)$$

$$\theta_2 = \sin^{-1}\left(\left(\frac{1.0001}{1.5}\right)\sin(30°)\right) = 19.47°$$

25.5.2.1 Speed of Light and Index of Refraction

The index of refraction was created to express the bending of a light ray as it crosses the interface between two different materials. This bending of a light ray is related to the speed of light in the different materials, so the index of refraction is the ratio of the speed of light in vacuum (c) to the speed of light in the material (v).

$$n = \frac{c}{v} \qquad (25.8)$$

Thus, in a vacuum $n = \frac{c}{v} = \frac{c}{c} = 1$.

Example

Compute the speed of light in glass.

Solution

Since the index of refraction in glass is 1.5, the speed of light in glass is,

$$v = \frac{c}{1.5} = 2 \times 10^8 \, \frac{m}{s}$$

This concept of the speed of light and the bending of the light leads us to think about the time it takes a light ray to travel from the source to where it is sensed. It turns out that both the law of reflection and Snell's Law of refraction can be proven mathematically by Fermat's Principle, which states that the time it takes light to travel from one location to another is always a minimum. Therefore, these empirical laws do have as one of their fundamental concepts a law of nature that sets a limit on a fundamental quantity like time. So, at its core optics is a set of rules that can be applied to predict the outcome of a physical situation, not unlike other rules like the conservation laws.

25.5.3 THE PRISM

A prism is a piece of glass, or other transparent material, cut into a specific geometry and polished. To understand the significance of the shape of a prism it is important to understand the path a light ray will take upon entering and exiting a cube of glass, as depicted in Figure 25.12.

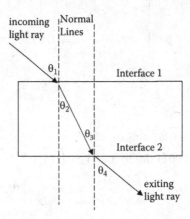

FIGURE 25.12 Refraction of light.

Assuming that the interfaces of air and glass are parallel, the angle the light enters the cube (θ_1) will equal the angle at which the light exits (θ_4) the cube of glass. Since the bending of the light at the first interface is equal and opposite the bending of the light ray at the second interface, the light ray is simply shifted over an amount that is proportional to the thickness of the glass, but parallel to the original path of the light ray.

The most common geometry for a prism is a triangle, as shown in Figure 25.13.

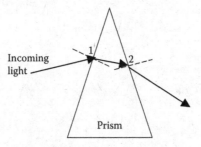

FIGURE 25.13 Bending of light with a prism.

To understand the refraction of light passing through a prism, it is best to study the interfaces at each face of the prism separately. As the incident light ray interfaces with face 1 of the prism, it is bent toward the normal (dashed line). The ray of light travels through the prism, and upon exiting at interface 2, the ray bends further away from the normal, since the index of refraction of air is less than the index or refraction of glass.

Light rays entering and exiting the prism are refracted. On the way in, light rays are bent closer to the normal because the light is moving from air to glass. Snell's Law tells us that as light moves from a material with a smaller index of refraction to one with a larger index, the angle of refraction must be less than the angle of incidence. On the way out, the rays are bent away from the normal for the opposite reason. Since the faces of the triangular prism are "tilted" in opposite directions, the entrance and exit refraction increase the overall bending of the light.

25.5.4 DISPERSION

The last concept needed to understand how a prism separates white light into all the colors of the spectrum is dispersion. The index of refraction describes the speed of light and the expected bending of a yellow light ray at a wavelength of $\lambda = 589$ nm in a material, such as a glass. Different colors of light bend different amounts, as shown in Figure 25.14, so they have slightly different indices of refraction.

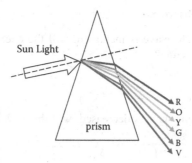

FIGURE 25.14 Spreading out the visible spectrum with a prism.

The difference in the index of refraction for different colors of light is called *dispersion*. Since violet light bends the most, it has a slightly higher index of refraction than the rest of the colors in the visible spectrum, and since red light bends the least, it has a slightly lower index of refraction than the rest of the colors in the visible spectrum. White light incident on the prism will be separated into its colors, because the red light will bend the least at each face and the violet light will bend the most.

25.6 ANSWER TO CHAPTER QUESTION

Today, we explain color as a property of light linked to the wavelength of light waves and the energy of the photons that make up the beam of light. A prism breaks a beam of light into its colors by refraction and dispersion. One of the first formal experimental studies of color with prisms was performed by Isaac Newton in the late 1600s. One of his most important experiments, depicted in Figure 25.15, involved separating a beam (ray) of sunlight into the visible spectrum of colors with a prism and then testing if those colors could be split apart again with a second prism.

Newton used a board with a hole in it to separate the individual colors of light rays before he put them through a second prism. Newton discovered that the light rays of color could not be broken up further, so he concluded that the prism spread the (white) sunlight into its colors as opposed to the belief at the time that the prism added the color to the sunlight.

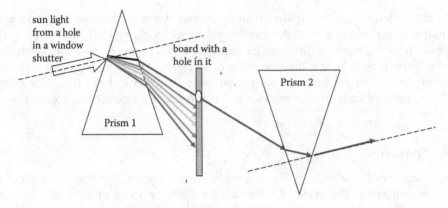

FIGURE 25.15 Newton's experiment to investigate color.

25.7 QUESTIONS AND PROBLEMS

25.7.1 MULTIPLE CHOICE QUESTIONS

1. If the frequency of a light wave is increased, will the energy of the photons increase, decrease, or remain the same?
   A. increase
   B. decrease
   C. remain the same
2. If the frequency of a light wave is decreased, will the wavelength of the light increase, decrease, or remain the same?
   A. increase
   B. decrease
   C. remain the same
3. Will light travel faster, slower, or the same speed through materials with larger indices of refraction as compared with those with a smaller index of refraction?
   A. faster
   B. slower
   C. at the same speed in all materials
4. Suppose you are seated in a small boat on a lake and that you are looking into the water at a fish that appears to be a couple of feet below the surface and a few feet east of where you are sitting. If you point directly at where the fish appears to be, you are pointing:
   A. above the fish
   B. below the fish
   C. straight at the fish
5. If the frequency of a light wave is increased, will the energy of the photons increase, decrease, or remain the same?
   A. increase
   B. decrease
   C. remain the same
6. If the frequency of a light wave is decreased, will the speed of the light increase, decrease, or remain the same?
   A. increase
   B. decrease
   C. remain the same

7. The surface of the sun is 5800 K, so its black body curve is centered on a wavelength in the yellow part of the spectrum. If the temperature of the sun cooled, would the center wavelength of the sun move toward the blue part of the spectrum, toward the red part of the spectrum, or stay at the same wavelength?
   A. toward blue
   B. toward red
   C. stay at the same wavelength

8. When magenta and yellow ink are mixed as a spot on a page, what color is observed?
   A. blue
   B. green
   C. red
   D. black

9. When blue and green pixels are illuminated on the TV screen, that spot on a screen is observed as what color?
   A. cyan
   B. yellow
   C. magenta
   D. black

10. When a yellow banana is illuminated with blue light, what color will the banana appear?
    A. cyan
    B. yellow
    C. magenta
    D. black

25.7.2 PROBLEMS

1. Compute the energy of a photon of light with a frequency of 1.2×10^{14} Hz.

2. Compute the energy of a photon of light with a frequency of 2.5×10^{14} Hz.

3. Compute the wavelength of light traveling through air that has a frequency of 1.2×10^{14} Hz.

4. Compute the wavelength of light traveling through air that has a frequency of 2.5×10^{14} Hz.

5. Compute the frequency of a light traveling through air that has a wavelength of 570 nm.

6. Compute the frequency of a light traveling through air that has a wavelength of 632 nm.

7. A light ray is incident from air at an angle of 30° measured from the surface of a calm swimming pool. At what angle does the refracted ray travel in the water?

8. A light ray is incident from air at an angle of 20° measured from the surface of a flat piece of glass. At what angle does the refracted ray travel in the glass?

9. A light ray in air is incident on an air-to-glass interface at an angle of 45° and is refracted at an angle of 30° to the normal. What is the index of refraction of the glass?

10. Compute the wavelength of light traveling through glass that has a frequency of 1.2×10^{14} Hz.

26 Geometric Optics

26.1 INTRODUCTION

Geometric optics employs the rules of straight-line geometry to predict the paths on light rays as they reflect off surfaces and refract as they pass from one material to another. In this chapter, the laws of reflection and refraction are used to describe and predict the optical effect of lenses and mirrors. Specifically, the thin lens equation and the magnification equation are derived, and the steps required to develop a ray diagram are explained. Examples using both these algebraic and graphical techniques to predict the size, location, and orientation of an image from a single and multiple lens system are provided. In addition, the analysis for mirror-based systems is also developed, along with additional examples. The chapter culminates by applying these expressions and techniques to the analysis of the human eye and the understanding of how eyeglasses are used to correct vision.

26.2 CHAPTER QUESTION

How do prescription eyeglasses correct someone's vision? This question will be answered at the end of the chapter utilizing the methods of geometric optics developed in the chapter.

26.3 INTRODUCTION TO LENSES

A lens is a piece of transparent material, often glass or acrylic, which is shaped to bend light in a specific manner. The two fundamental types of lenses are converging and diverging. A converging lens, sketched in Figure 26.1, focuses rays that are parallel to each other and perpendicular to the face of the lens to a point in front of the lens. On the other hand, a diverging lens spreads parallel rays that are perpendicular to the face lens so that they diverge from a point in back of the lens. The distance from the lens to the point of convergence or divergence is the focal length of the lens. This is demonstrated in Figures 26.1 and 26.2 for converging and diverging lenses, respectively.

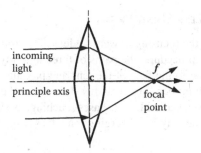

FIGURE 26.1 Side view of a converging lens.

The focal point for a lens is the place the rays cross for a converging lens. For a converging lens, it is the point from which the rays originated if they were traveling in the path they are after the lens. The distance from the center of the lens, **c**, to the focal point is called focal length (f). To indicate that the focal point of the two different types of lenses are on opposite sides, it was chosen to assign a positive to the focal length of the converging lenses ($f+$) and a negative to the focal

DOI: 10.1201/9781003308072-27

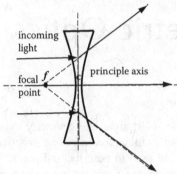

FIGURE 26.2   Side view of a diverging lens.

length of the diverging lenses ($f-$). Thus, it is common to refer to converging lenses as positive lenses and diverging lenses as negative lenses.

The diameter (D) of the lens is measured across the face of the lens, as demonstrated in Figure 26.3.

FIGURE 26.3   Front view of a lens.

Together, the focal length (f) and diameter (D) of the lens describe the speed of the lens. The f-number ($f\#$) of a lens is the quotient of the focal length of the lens and the diameter of the lens, given in equation (26.1) as

$$f\# = \frac{f}{D} \tag{26.1}$$

The smaller the $f\#$, the faster the lens, since more light is brought to a focus closer to the lens, and conversely, the larger the $f\#$, the slower the lens. This is helpful in photography since the slower the lens the longer the exposure time that is needed to create an image.

26.3.1   DESCRIPTION OF A LENS USING PRISMS

The easiest way to understand the focusing power of a lens is to "reflect" upon the bending of light by a prism. To understand this procedure, the paths of three light rays must be understood. First, a light ray traveling perpendicular to the face of the lens and passing through the center of the lens, will pass straight through with no bending. This ray determines the principle axis of the lens.

If two identical prisms are stacked with their bases touching, as shown in Figure 26.4, they will bend incoming light in the same way a converging lens does. Light rays traveling parallel to the

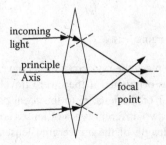

FIGURE 26.4   Stacked prisms, base to base, make a converging lens.

principle axis of the lens will strike the lens so that each face of the lens causes a bending that adds together to result in a net bending of the ray toward the principle axis. Since the top and bottom prism are identical, light rays from each side will cross at a point on the principle axis denoted as the focal point of the lens.

If two identical prisms are stacked with their tips touching, as shown in Figure 26.5, they will bend incoming light in the same way a diverging lens does. Light rays traveling parallel to the principle axis of the lens will strike the lens so that each face of the lens causes a bending that adds together to result in a net bending of the ray away from the principle axis. Since the top and bottom prisms are identical, light rays from each side will diverge away from a point on the principle axis denoted as the focal point of the lens.

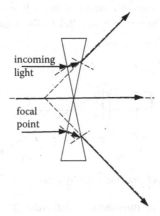

FIGURE 26.5   Stacked prisms, point to point, make a diverging lens.

26.3.2 OPTICAL IMAGES

One of the most common applications of lenses is to produce images with devices such as projectors. There are two types of images: real and virtual. A real image is the one you see every time you go to the movies or look at an overhead projection or the projection of a computer screen in a classroom. A virtual image may not be as obvious to you, but you may have seen one if you have ever used a magnifying glass. The object you are looking at appears enlarged, but this image is constructed in your brain and cannot be projected onto a screen. The simplest way to determine if an image produced by a particular lens arrangement will be real or virtual is with a ray diagram.

26.4 RAY DIAGRAMS

A ray diagram is a sketch based on a few simple rules that provides information about the location, size, and orientation of the image produced in a specific optical arrangement.

To make a ray diagram use the following steps:

1. Draw a straight horizontal line across the paper and label it as the principle axis, as shown in Figure 26.6.

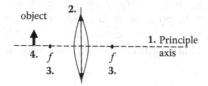

FIGURE 26.6   Steps 1–4 for drawing a ray diagram.

2. Locate the position of the lens on your sketch with a straight vertical line that intersects the center of principle axis. You don't need to sketch the lens as shown in Figure 26.6, but it is sometimes helpful so you remember the type of lens, converging in this case, that is being analyzed.
3. Locate the focal points, f, of the lens or mirror on the principle axis. Remember that these focal points should be the same distance from either side of the lens.
4. Sketch a short vertical arrow on the left side of the principle axis at the location of the object. Be sure to locate the object a distance away from the lens relative to the focal length. For example, if a converging lens with a focal length of 5 cm has an object located 7 cm away from the lens, make sure that the object is the correct relative distance away from the lens. Notice in Figure 26.6 that the vertical arrow is located seven gaps in the dashed line to the left of the lens and the focal point is five gaps in the same dashed line.
5. As shown in Figure 26.7, draw a line parallel to the principle axis from the tip of the object arrow to the lens line.

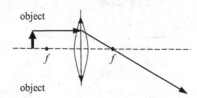

FIGURE 26.7   Step 5 in the process of drawing a ray diagram.

From the point where the parallel line intersects the lens line, draw a line through the focal point on the right side of the ray diagram.

6. As shown in Figure 26.8, draw another line starting from the tip of the object arrow going through the focal point on the left of the ray diagram and stopping at the lens line.

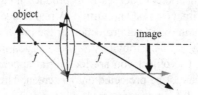

FIGURE 26.8   Step 6 in the process of drawing a ray diagram.

From the intersection point of the line you drew and the lens line, draw a line parallel to the principle axis. The point at which the two lines on the right side of the lens intersect is the location of the top of the image.

In the example used to describe the ray diagram process in the previous section, an object was placed to the left of a converging lens at a distance greater than the focal length, and an inverted image is formed on the right-hand side of the lens where the rays crossed. The image at this location of crossed rays is referred to as a *real image*, because the image at the location predicted by the ray diagram can be projected onto a screen. The image exists if there is a person looking at it or not. This is the type of image that a movie projector generates at the theater. You may have noticed that the film slides past the projector bulb up-side-down, so that the image it produces is right-side-up. The image, as is the screen, is located at the points of intersection of light rays. When the image is out of focus, the light is still incident on the screen; it is just not at the intersection point of the light rays.

26.4.1 Real Image from a Converging Lens
Example

An object is placed 15 cm to the left of the lens, with a focal length of 10 cm. Use rays to find the location of the image and characteristics of the image.

Solution

Sketch the ray diagram following the steps outlined above. From the diagram below, the image is real, larger, inverted, and further away from the lens than the object. The approximate distance of the image from the center of the lens is 30 cm.

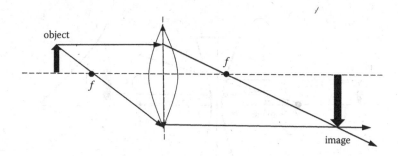

FIGURE 26.9 Ray diagram of an object outside the focal length of a converging lens.

26.4.2 Virtual Images from a Converging Lens

The second type of image that can be produced by a converging lens is a virtual image. Unlike a real image, which can be projected, a virtual image is produced in the mind of the observer. This is why it is called *virtual*. If you have ever used a magnifying glass to see a small object, like an ant, you have produced a virtual image in your mind.

The major difference in the way that virtual and real images are produced, is that a virtual image comes from diverging light rays. As demonstrated in Figure 26.10, the ray coming from the head of the ant, which is parallel to the principle axis, passes through the focal point on the opposite side of the lens. The ray coming from the head of the ant and in a direction from the focal length on the same side of the lens as the ant, diverges parallel to the principle axis. The two rays on the right

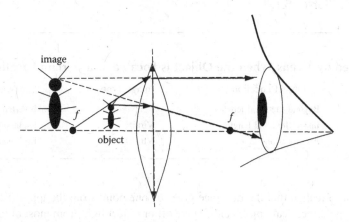

FIGURE 26.10 Ray diagram of an object inside the focal length of a converging lens.

side of the lens and entering the eye, diverge. They will never cross, so they cannot form a real image. The observer uses their mind to trace back through the lines of sight (the dashed lines in Figure 26.6), to generate a virtual image, which is larger than the object and still oriented in the same way as the object.

26.4.3 Virtual Image Produced by a Diverging Lens

As described in the first section of the chapter and shown in Figure 26.11, a diverging lens is wider at the edges and thinnest in the middle. As its name implies, rays of light diverge from a point as they pass through a diverging lens.

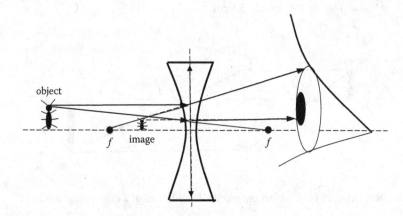

FIGURE 26.11 Ray diagram of an image produced with a diverging lens.

A ray diagram for this lens follows similar rules, except that the light ray starting at the top of the object and traveling parallel to the principle axis, diverges away from the lens at an angle away from the focal point on the same side of the lens as the object. The light ray traveling in a direction that would take it through the focal point on the other side of the lens is bent so that it is parallel to the principle axis. Since the two rays will not converge, the observer must trace back through the lines of sight (dashed lines in Figure 26.11) to produce a virtual image.

For a converging lens, the image is always smaller than and oriented in the same way as the object.

The types of images produced when an object is located in a particular place in front of a lens are summarized in Table 26.1.

TABLE 26.1

Images Produced by a Lens When the Object is Located at a Certain Location

Lens Type	Object Location	Image Type	Image Characteristics
Converging	Beyond the focal length	Real	Inverted (larger or smaller)
Converging	Inside the focal length	Virtual	Right-side-up larger
Diverging	Any location	Virtual	Right-side-up smaller

It is important to realize that the ray trace gives us one point from the top of the object to find the top of the image. Since real objects either give off or reflect light from most of their surface and light rays leave every point of every object in all directions, the image we see is the combination of many ray traces starting from every point on the object. The location of the image is not a point but a plane, which is called the *focal plane*.

26.5 THIN LENS AND MAGNIFICATION FORMULAS

The thin lens and magnification formula can be used instead of, or in conjunction with, a ray diagram to predict the location and size of the image produced by a lens. As described in Figure 26.12, the object distance (d_o) is measured from the center of the lens to the object, and the image distance (d_i) is measured from the center of the lens to the image. Both are the absolute value of the distance.

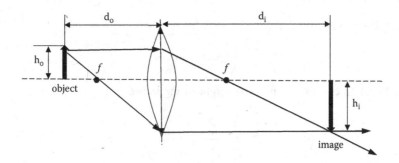

FIGURE 26.12 The thin lens formula and the ray diagram labels.

The absolute value of the heights of the object and image are labeled as h_o and h_i, respectively.
The Thin Lens Formula, which relates the object distance to the image distance and the focal length of the lens, is given in equation (26.2) as

$$\frac{1}{f} = \frac{1}{d_o} + \frac{1}{d_i} \tag{26.2}$$

The magnification (m) of the lens is defined as the ratio of height of the image (h_i) to the height of the object (h_o). The magnification formula is given in equation (26.3) as

$$m = \frac{h_i}{h_o} = -\frac{d_i}{d_o} \tag{26.3}$$

The thin lens formula relates the location of the object to the location of the image through the focal length of the lens. These two expressions, the thin lens equation and the magnification equation, can be applied to find the size and location of an image given location and size of the object and the information about the lens.

Example

Calculate the location and size of the image of a 3 cm tall object, which is located 6 cm in front of a converging lens, with a focal length of 4 cm.

Solution

Solve the thin lens equation for d_i:

$$\frac{1}{d_i} = \frac{1}{f} - \frac{1}{d_o}$$

Plug the known quantities into the thin lens equation:

$$\frac{1}{d_i} = \frac{1}{4 \text{ cm}} - \frac{1}{6 \text{ cm}}$$

Use common denominators or other methods to solve for d_i:

$$\frac{1}{d_i} = \frac{3}{12 \text{ cm}} - \frac{2}{12 \text{ cm}} = \frac{1}{12 \text{ cm}}$$

So, the image distance is:

$$d_i = 12 \text{ cm}$$

Use the magnification equation to find the height of the image:

$$m = \frac{h_i}{h_o} = -\frac{d_i}{d_o}$$

So, the magnification of this lens for this image is:

$$m = -\frac{12 \text{ cm}}{6 \text{ cm}} = -2$$

Since the magnification is also equal to the ratio of the heights:

$$h_i = m \, h_o = (-2)(3 \text{ cm}) = -6 \text{ cm}$$

Notice that the image height is negative, which indicates the image is inverted, up-side-down.

There are a few rules to remember when using the thin lens and magnification formulas.

1. Converging lenses have positive focal lengths.
2. Diverging lenses have negative focal lengths.
3. A positive image distance indicates a real image.
4. A negative image distance indicates a virtual image.
5. A positive magnification indicates an image with the same orientation as the object.
6. A negative magnification indicates an image with the opposite orientation as the object.

26.5.1 Derivation of the Thin Lens and Magnification Equations

The thin lens equation and the magnification equation can be applied without understanding exactly how they are derived, but the following derivation of these expressions may shed some light on the underlying structure of the subject. This derivation is completely geometric and starts with Figure 26.13.

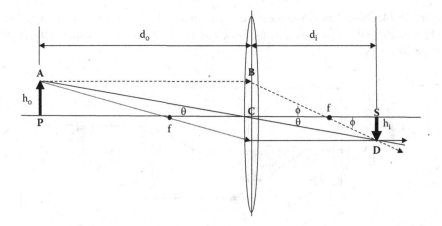

FIGURE 26.13   Diagram of the geometry needed for the derivation of the thin lens and magnification formulas.

From triangle (ACP): $\tan(\theta) = \frac{h_o}{d_o}$

From triangle (DCS): $\tan(\theta) = \frac{h_i}{d_i}$

From triangle (BCf): $\tan(\phi) = \frac{h_o}{f}$

From triangle (DSf): $\tan(\phi) = \frac{h_i}{(d_i - f)}$

From triangles ACP and DCS: $\frac{h_o}{d_o} = \frac{h_i}{d_i}$

From triangles BCf and DSf: $\frac{h_o}{f} = \frac{h_i}{(d_i - f)}$

Thus, $\frac{d_o}{d_i} = \frac{f}{(d_i - f)}$ which can be written as the thin lens in equation (26.2)

$$\frac{1}{f} = \frac{1}{d_o} + \frac{1}{d_i}$$

The magnification (m) is defined as the ratio of the height of the image to the height of the object. From triangles ACP and DCS: $\frac{h_o}{d_o} = \frac{h_i}{d_i}$, so $m = \frac{h_i}{h_a} = \frac{d_i}{d_o}$.

Because these expressions are derived for the situation described in Figure 26.9, it is common to input the negative sign, since the image is inverted as compared to the image, so the magnification equation is commonly written as shown in equation (26.3) as

$$m = \frac{h_i}{h_o} = -\frac{d_i}{d_o}$$

These fundamental expressions of the geometric optics are derivations applying trigonometric functions and rules of geometry.

26.6   IMAGE RELAY

Image relay is the technique used when analyzing an optical system made up of two or more lenses. The fundamental concept behind image relay is that the image produced by the first lens becomes the object that is imaged by the second lens, and so on. The image you see when looking through a compound microscope is produced in this way.

In Figure 26.14, object 1 is a distance d_{o1} away from lens 1, with a focal length of f_1. The image of this object is located a distance d_{i1} away from lens 1, which can be computed with the thin lens equation for this lens.

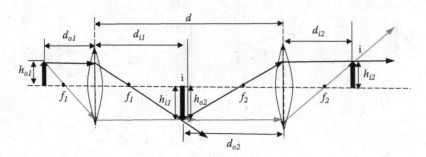

FIGURE 26.14   Image relay.

Since lens 1 is a distance d away from lens 2 and image 1 is to be considered the object for lens 2, the object distance for the second lens is $d_{o2} = (d - d_{i1})$. The image distance from the second lens (d_{i2}) can be found with the thin lens equation for this lens.

It may be easier to follow the procedure of image relay with a numerical example, with numbers that correspond closely to the diagram in Figure 26.14.

26.6.1   EXAMPLE OF IMAGE RELAY

Two converging lenses, with focal lengths of 6 cm and 4 cm, are set 24 cm apart along a common principle axis. An object is placed 9 cm in front of the 6 cm focal length lens. What is the location of the image after the 4 cm focal length lens, and what is the total magnification of the system?

Solution

Label the known values: $f_1 = 6$ cm, $f_2 = 4$ cm, $d = 26$ cm, and $d_{o1} = 9$ cm.

Solve the thin lens equation for d_i:

$$\frac{1}{d_{i1}} = \frac{1}{f_1} - \frac{1}{d_{o1}} = \frac{1}{6 \text{ cm}} - \frac{1}{9 \text{ cm}} = \frac{3}{18 \text{ cm}} - \frac{2}{18 \text{ cm}} = \frac{1}{18 \text{ cm}}$$

So, $d_{i1} = 18$ cm

Find $d_{o2} = (d - d_{i1}) = (26 \text{ cm} - 18 \text{ cm}) = 8$ cm

Solve the thin lens equation for d_{i2}:

$$\frac{1}{d_{i2}} = \frac{1}{f_2} - \frac{1}{d_{o2}} = \frac{1}{4 \text{ cm}} - \frac{1}{8 \text{ cm}} = \frac{2}{8 \text{ cm}} - \frac{1}{8 \text{ cm}} = \frac{1}{8 \text{ cm}}$$

So, $d_{i2} = 8$ cm.

To find the total magnification of the system, find the magnification for each lens system and multiply the magnifications together.

$$m_1 = -\frac{d_{i1}}{d_{o1}} = -\frac{18 \text{ cm}}{9 \text{ cm}} = -2$$

$$m_2 = -\frac{d_{i2}}{d_{o2}} = -\frac{8 \text{ cm}}{6 \text{ cm}} = -\frac{4}{3}$$

$$m = m_1 m_2 = (-2)\left(-\frac{4}{3}\right) = \frac{8}{3}$$

26.6.2 EXAMPLE OF A COMPOUND MICROSCOPE

A compound microscope is made with two converging lenses. The lens closest to the eye is the eyepiece, and the lens closest to the object is the objective, as shown in Figure 26.15.

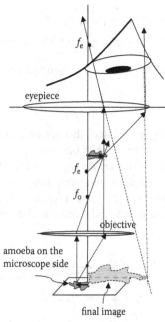

FIGURE 26.15 Compound microscope.

For the microscope in this example, the eyepiece has a focal length of 10 cm, the objective as a focal length of 4.5 cm, and the eyepiece is 20 cm away from the objective. The microscope slide with an amoeba on it is a distance of 7 cm from the objective and a distance of 27 cm from the eyepiece. Compute the total magnification of this compound microscope with the lenses and objects at the location specified. Figure 26.15 is not to scale but was constructed so that a ray trace of the situation could be constructed. The objective lens produces a real image of the amoeba that is observed through the eyepiece, and a virtual image is produced in the mind of the observer.

Solution

Label the known values: f_o = 4.5 cm, f_e = 10 cm, d = 20 cm, and d_{o1} = 7 cm.
   Solve the thin lens equation for d_{i1}:

$$\frac{1}{d_{io}} = \frac{1}{f_o} - \frac{1}{d_{oo}} = \frac{1}{4.5 \text{ cm}} - \frac{1}{7 \text{ cm}}$$

So, d_{io} = 12.6 cm
   Find $d_{oe} = (d - d_{io}) = (20 \text{ cm} - 12.6 \text{ cm}) = 7.4$ cm
   Solve the thin lens equation for d_{i2}:

$$\frac{1}{d_{ie}} = \frac{1}{f_e} - \frac{1}{d_{oe}} = \frac{1}{10 \text{ cm}} - \frac{1}{7.4 \text{ cm}}$$

So, d_{ie} = −28.5 cm; therefore, it is virtual.

To find the total magnification of the system, find the magnification for each lens system and multiply the magnifications together.

$$m_o = -\frac{d_{io}}{d_{oo}} = -\frac{12.6 \text{ cm}}{7 \text{ cm}} = -1.8$$

$$m_e = -\frac{d_{ie}}{d_{oe}} = -\frac{-28.5 \text{ cm}}{7.4 \text{ cm}} = +3.85$$

$$m = m_1 m_2 = (-1.8)(+3.85) = -6.93$$

Therefore, this compound microscope provides a magnification of about 7x.

26.7 MIRRORS

Like lenses, mirrors are important optical devices that are all around us. The most common type of mirror is the flat mirror that hangs on the walls in our homes. The image we see in a flat mirror is virtual, since the rays reflected off the mirror are diverging and we produce the image in our brain. As demonstrated in Figure 26.16, the image of ourselves that we see in a flat mirror is traced back through the lines of sight, similar to the virtual images produced by lenses. The main difference is that light rays from our face reflect off the mirror and reach our eyes instead of passing through the lens.

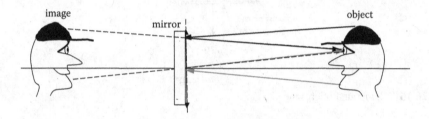

FIGURE 26.16 Ray diagram of an image produced with a flat mirror.

The most common curved mirror is the one on the side view mirrors on our car. These convex mirrors are good, since they allow us a wider field of view, but they also make objects appear smaller. This explains the warning, "objects may be closer than they appear", on the side view mirror. As demonstrated in Figure 26.17. the image of ourselves that we see in a convex mirror is traced back through the lines of sight, similar to the virtual images produced by lenses.

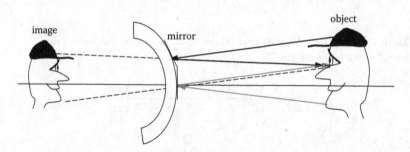

FIGURE 26.17 Ray diagram of a reflection produced with a convex mirror.

The ray that starts at the chin of the observer and reflects just like it does off a flat mirror, since at the center of the curve the tangent is perpendicular to the principle axis. On the other hand, the ray from the observer's hat reflects closer to the center of the convex mirror than the ray starting from the same point for the flat mirror reflection. Upon tracing back through the line of sight, it is clear that the image will be smaller.

The concave mirror, which has its reflective coating on the inside of the curve, is another curved mirror, which you may have used. These are commonly used as a make-up mirror to magnify your image. As demonstrated in Figure 26.18, the image of ourselves that we see in a concave mirror is traced back through the lines of sight.

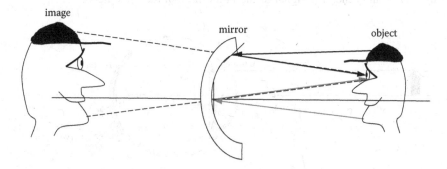

FIGURE 26.18   Ray diagram of a reflection produced with a concave mirror.

The ray that starts at the chin of the observer and reflects just like it does off a flat mirror, since at the center of the curve the tangent is perpendicular to the principle axis. On the other hand, the ray that starts at the top of the hat reflects further from the center of the concave mirror than the ray starting from the same point for the flat mirror reflection. Upon tracing back through the line of sight, it is clear that the image will be larger.

These concave mirrors can also produce real images. One of the main uses of concave mirrors is to produce real images of astronomical objects as the primary mirror of reflecting telescopes. Many telescopes produce real images that are recorded on digital devices. You don't need to look through the optics of many telescopes to see the images; therefore, they must be real. The ray diagram procedure and thin lens and magnification formulas can be used with mirrors with some slight adjustments.

The most obvious adjustment to the ray diagram procedure, is that the rays reflect off the surface of the mirror. The ray parallel to the principle axis is reflected through the focal point, and the ray passing through the focal point reflects parallel to the principle axis. At the spot where the rays cross, the image is produced. In Figure 26.19, the real image is smaller and inverted as compared to the object.

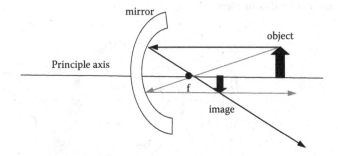

FIGURE 26.19   Ray diagram of a real image produced with a concave mirror.

The thin lens equation can also be used to predict the location of the image by using a positive focal length for concave mirrors and a negative focal length for convex mirrors. It may now be obvious that concave mirrors focus light like converging lenses and convex mirrors diverge light from a point like diverging lenses. Thus, these devices bend light in very similar ways.

26.8   ANSWER TO CHAPTER QUESTION

To understand how eyeglasses help correct vision, you must first understand how the eye generates an image. As described in Figure 26.20, the light rays from an object are focused by the optical power of the combination of the cornea and eye lens. A real image is produced on the retina, which encodes the image and sends the signal to the brain through the optic nerve.

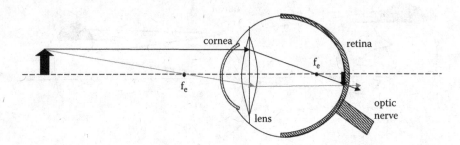

FIGURE 26.20   Ray trace of the eye.

Someone who is farsighted (hyperopic) has an eyeball that is too short for their lens system (Figure 26.21), so the location of the real image produced by the cornea lens combination is behind the eye.

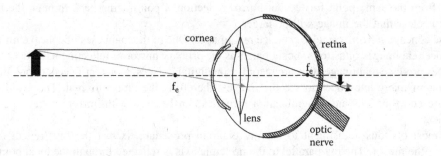

FIGURE 26.21   Ray trace of a farsighted eye.

The solution is therefore a positive lens, as shown in Figure 26.22, which will bring the image to a focus on the retina and results in clear vision.

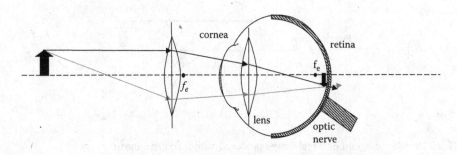

FIGURE 26.22   Ray trace of a positive lens providing correction for a farsighted eye.

To find the correct lens, an image relay calculation can be done with a standard object, like a letter on an eye chart. You can tell if someone is farsighted if you look at the side of their face through their glasses. If they have a positive lens in their glasses the side of their face will be within the focal length of the lens, so to you it will act as a magnifying glass and the side of their face will be expanded beyond the normal shape of their face.

Someone who is nearsighted, myopic, has an eyeball that is elongated (Figure 26.23), so the location of the real image produced by the cornea lens combination is in front of the retina as described in Figure 26.23.

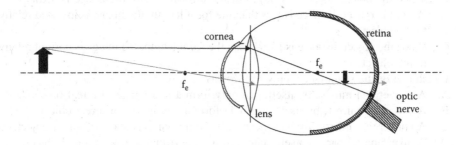

FIGURE 26.23   Ray trace of a nearsighted eye.

The solution is therefore a negative lens, as demonstrated in Figure 26.24, which will bring the image to a spot on the retina.

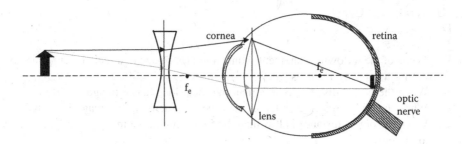

FIGURE 26.24   Ray trace of the correction of a nearsighted eye.

Today, a popular solution to these problems is laser eye correction surgery. In this procedure, the cornea is reshaped with laser light to correct the focus point of the eye. So, instead of putting a lens in front of the eye to change the path of the incoming light, the focus of the eye system is changed so the image is on the retina.

26.9   QUESTIONS AND PROBLEMS

26.9.1   MULTIPLE CHOICE QUESTIONS

1. Which one of the statements below is correct?
    A.   A converging lens, by itself, can only form a real image of a real object.
    B.   A diverging lens, by itself, can only form a real image of a real object.
    C.   A converging lens, by itself, can only form a virtual image of a real object.
    D.   A diverging lens, by itself, can only form a virtual image of a real object.
2. A converging lens of focal length f produces, by itself, an image of a real object that is smaller than the object. The relationship between object distance d_o and the focal length must be:
    A. $d_o > 2f$
    B. $f < d_o < 2f$

C. $d_o = f$

D. $f > d_o > .5f$

E. $d_o < .5f$

3. Which one of the following statements concerning the image of a real object formed by a converging lens, by itself, is true?
   A. When the object distance is less than the focal length, the image is virtual.
   B. When the object distance is larger than the focal length, the image is virtual.
   C. When the object distance is less than the focal length, the image is inverted relative to the object.
   D. When the object distance is larger than the focal length, the image is upright relative to the object.

4. Which one of the statements below is correct?
   A. A converging mirror, by itself, can only form a real image of a real object.
   B. A diverging mirror, by itself, can only form a real image of a real object.
   C. A converging mirror, by itself, can only form a virtual image of a real object.
   D. A diverging mirror, by itself, can only form a virtual image of a real object.

5. A converging mirror of focal length f produces, by itself, an image of a real object that is smaller than the object. The relationship between object distance d_o and the focal length must be:
   A. $d_o > 2f$
   B. $f < d_o < 2f$
   C. $d_o = f$
   D. $f > d_o > .5f$
   E. $d_o < .5f$

6. Which one of the following statements concerning the image of a real object formed by a converging mirror, by itself, is true?
   A. When the object distance is less than the focal length, the image is virtual.
   B. When the object distance is larger than the focal length, the image is virtual.
   C. When the object distance is less than the focal length, the image is inverted relative to the object.
   D. When the object distance is larger than the focal length, the image is upright relative to the object.

7. Which one of the following statements is most correct?
   A. A convex lens is a converging lens, and a convex mirror is a converging mirror.
   B. A concave lens is a diverging lens, and a concave mirror is a diverging mirror.
   C. A convex lens is a converging lens, whereas a convex mirror is a diverging mirror.
   D. A concave lens is a diverging lens, whereas a concave mirror is a converging mirror.
   E. Both a and b above are correct.
   F. Both c and d above are correct.
   G. None of the above are correct.

8. What kind of lens is a magnifying glass?
   A. converging
   B. diverging

9. What does it mean for the object distance h_o to be negative?
   A. It doesn't mean anything; an object distance can never be negative.
   B. It can only happen in an optical system consisting of more than one optical element. More specifically, it can only happen if the object is the image of another optical element. It means that the object is on the side of the lens opposite from where the light is coming. So, for instance, if the light is traveling left to right, the object distance is negative when the object is to the right of the optical element.

10. A 1.0 cm tall object is placed 8.0 cm in front of a converging lens with a focal length of magnitude $|f| = 2.0$ cm. Is the image larger than the object, is it smaller than the object, or is it the same size as the object?

A. larger

B. smaller

C. the same size

26.9.2 PROBLEMS

1. An object is positioned 14 cm in front of a lens. The image produced is 5.8 cm behind the lens. Find the focal length of the lens.

2. For an object of height 7 cm that is 12 cm from the plane of a lens having a focal length of 5 cm, draw a ray diagram of the arrangement. Based on your ray diagram, state if the image is real or virtual, right-side-up or up-side-down, larger or smaller than the object.

3. For an object of height 7 cm that is 12 cm from the plane of a lens having a focal length of 5 cm, calculate the image distance, the magnification, and the image height.

4. For an object of height 8.5 cm that is 11 cm from the plane of a diverging lens whose focal points are 6 cm from the plane of the lens, draw a ray diagram of the arrangement. Based on your ray diagram, state if the image is real or virtual, right-side-up or up-side-down, larger or smaller than the object.

5. For an object of height 8.5 cm that is 11 cm from the plane of a diverging lens whose focal points are 6 cm from the plane of the lens, calculate the image distance, the magnification, and the image height.

6. An object is positioned 25 cm in front of a mirror. The image produced is 15 cm in front the mirror. Find the focal length of the mirror.

7. For an object of height 6 cm that is 6 cm from the plane of a diverging mirror whose focal points are 12 cm from the plane of the mirror, draw a ray diagram of the arrangement. Based on your ray diagram, state if the image is real or virtual, right-side-up or up-side-down, larger or smaller than the object.

8. For an object of height 6 cm that is 6 cm from the plane of a diverging mirror whose focal points are 12 cm from the plane of the mirror, calculate the image distance, the magnification, and the image height.

9. Two lenses are arranged to produce a good-quality image of an object on a screen. For one particular arrangement, a converging lens (lens 1) with a focal length of 2 cm is located 8 cm in front of a converging lens (lens 2) with a focal length of 1 cm. An object, with a height of 5 cm, is located 10 cm in front of lens 1, and a clear image is located on a screen, which is to the right of lens 2. Compute the location of the final image relative to lens 2 and the total magnification of the two-lens system.

10. A camera lens system is made up of a combination of two lenses to produce a good-quality image of an object on the image sensor. For one particular camera, a diverging lens with a focal length of −10 cm is located 7 cm in front of a converging lens with a focal length of 5 cm. A flower, with a height of 12 cm, is located 50 cm in front of the diverging lens, and a clear image is located on the image sensor, which is to the right of lens 2. Compute the location of the final image relative to lens 2 and the total magnification of the two-lens system.

27 Physical Optics

27.1 INTRODUCTION

Physical optics is the study of the wave-nature of light. The chapter begins with an introduction to the electromagnetic spectrum and a review of the properties of waves, first discussed in Chapter 15 of Volume 1 of this textbook. The relationship between the wavelength, frequency, and speed of an electromagnetic wave is discussed, and examples are provided across the electromagnetic spectrum. The wave properties of diffraction and interference are examined in the context of Young's double slit experiment. Polarization is also presented as another wave property of light, and examples of polarization by reflection and transmission are discussed. As a final example of the electromagnetic spectrum and the wave properties of all "light", the X-ray diffraction of the DNA molecule is presented.

27.2 CHAPTER QUESTION

How was the double-helix structure of DNA discovered? This question will be answered at the end of the chapter after the wave-nature of light is discussed.

27.3 ELECTROMAGNETIC WAVES

In Chapter 25 on light and color, the characteristics of color and brightness were explained, in part, as human observations of the frequency and amplitude of the light waves, respectively. It is important to note that the visible spectrum, which ranges from wavelengths from 400 nm to 700 nm, is only a small part of the possible range of wavelengths of these waves. As a charged particle, like an electron, oscillates, the electric field sensed by an observer changes, as described by the vertical arrows and the curve with the E-label in Figure 27.1.

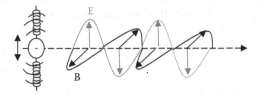

FIGURE 27.1 A vibrating charged particle generating an electromagnetic wave.

The moving charge is also a changing electric current, which generates a changing magnetic field, as described by the horizontal arrows and the curve with the B-label in Figure 27.1. The electric and magnetic fields are perpendicular to each other as the light wave moves through a vacuum. These concepts were grouped together and formalized in 1862 by James Maxwell. Today, it is believed that changing electric fields generate a magnetic field, which is Ampere's Law with Maxwell's Addition. It is also believed that a changing magnetic field generates an electric field, which is Maxwell's version of Faraday's Law. Therefore, once started by a vibrating charge, these electromagnetic waves are self-propagating. This is how light generated by stars millions of years ago travels across the universe for us to see. This self-propagating wave of vibrating electric and magnetic fields is known as an *electromagnetic wave,* or an *e&m wave,* for short. Waves of this type are known as radiation. Maxwell's work is one of the greatest discoveries of science that makes much of the way we communicate today possible.

DOI: 10.1201/9781003308072-28

27.3.1 Review of the Properties of Waves

The amplitude (A) of a wave is the maximum displacement from equilibrium, and the length of time from one peak to another is the period (T). In the case of an electromagnetic wave, the amplitude of the electric field is measured in Newtons per Coulomb and the period in seconds, usually with a prefix since the periods are extremely short times. Figure 27.2 describes the electric field part of an electromagnetic wave as that passes by on location over a time interval. The electric field of the e&m wave at one location will increase and decrease and change directions as time progresses. Depending on the type of the e&m wave, the electric field can be detected with different devices.

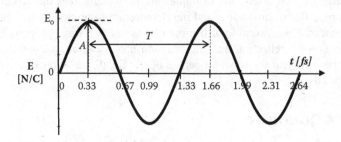

FIGURE 27.2   The displacement of an electromagnetic wave measured as a function of time.

For the wave in Figure 27.2, the amplitude is the maximum displacement from equilibrium and is denoted with an A and is a value of E_o, which is a standard variable for the magnitude of the electric field. For the wave in Figure 27.2, the period of the wave is the time it takes the wave to repeat, in this case from peak to peak, and is denoted with a T. The value of the period of this wave is

$$T = (1.66 - 0.33)fs = 1.33 \ fs = 1.33 \times 10^{-15}s$$

Another temporal-related variable is frequency for e&m waves, which is denoted with the symbol v and is related to period through a reciprocal relationship given in equation (27.1) as

$$v = \frac{1}{T} \tag{27.1}$$

So, for the wave in Figure 27.2, the frequency of the wave is

$$v = \frac{1}{1.33 \times 10^{-15}s} = 7.5 \times 10^{14} \text{ Hz}$$

Another way to observe a wave is to take a snapshot of the wave at one instant of time. If such an observation is made, a graph of the vertical components of e&m wave's displacement can be made vs. the horizontal distance. For this type of observation, a graph of displacement vs. time similar to Figure 27.3 is produced.

The amplitude of the wave is still the maximum displacement from equilibrium. For the wave in Figure 27.3, the wavelength of the wave is from peak to peak and is denoted with λ. The value of the period of this wave is

$$\lambda = (500 - 100) \text{ nm} = 400 \ \text{nm} = 400 \times 10^{-9} \text{ m} = 4.00 \times 10^{-7} \text{ m}$$

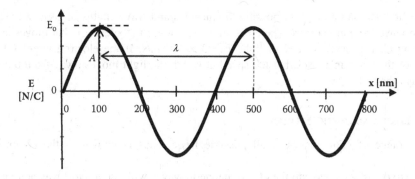

FIGURE 27.3   The displacement of an e&m wave measured as a function of distance.

For any wave, including e&m waves, the speed of the wave is related to the wavelength and frequency of the wave through the expression given in equation (27.2) as

$$c = \lambda \nu \qquad (27.2)$$

So, for the wave in Figures 27.2 and 27.3, the speed of the wave is

$$c = \lambda \nu = (4.00 \times 10^{-7} \text{ m})(7.5 \times 10^{14} \text{ Hz}) = 3 \times 10^{8} \frac{\text{m}}{\text{s}}$$

which is the speed of light in a vacuum.

The speed of light in a vacuum is given the variable name of **c** and has a specific value of **299,792,458 m/s**. As previously described, Maxwell discovered that any type of radiation, visible light included, is an electromagnetic wave. He discovered that the speed of light in a medium is related to the electric and magnetic properties of the medium. For a vacuum, the electric constant is the permittivity of free space, ε_0, which has a value of **$8.85418782 \times 10^{-12} \text{ m}^{-3} \text{ kg}^{-1} \text{ s}^4 \text{ A}^2$**, and the magnetic constant is the permeability of free space, μ_0, which has a value of **$1.25663706 \times 10^{-6} \text{ m}$ kg s^{-2} A^{-2}**. The permittivity of free space, ε_0, is the constant of proportionality in Gauss' Law between the charge enclosed by an area and the electric field. The constant for the magnetic field in a vacuum is the permeability of free space, μ_0, which is the value used in Ampere's Law to find the magnetic field due to an electrical current. If these values for ε_0 and μ_0 are plugged into equation (27.3)

$$c = \frac{1}{\sqrt{\varepsilon_0 \mu_0}} \qquad (27.3)$$

the resulting value is 299,792,458 m/s = c = speed of light. The units even work out: $1/[(\text{m}^{-3} \text{ kg}^{-1} \text{ s}^4 \text{ A}^2)(\text{m kg s}^{-2} \text{ A}^{-2})]^{1/2} = 1/[\text{s}^2/\text{m}^2]^{1/2} = \text{m/s}$. Maxwell arrived at equation (27.3) by finding the wave equation associated with the electric and magnetic fields of the e&m wave.

As mentioned before, the source of all electromagnetic waves is accelerating charged particles. The most common type of acceleration of charged particles that generate light, is the acceleration associated with vibrations. If a charged particle is vibrating like the mass on the end of a spring, the electromagnetic wave emanates outward from the vibrating charged particle like ripples on a pond. One of the important conceptual differences between water waves and an electromagnetic wave is that the e&m waves oscillate in 3D instead of the 2D surface of the water. So, if an antenna of a radio station is at the center of the oscillation, an electromagnetic wave will radiate out in a symmetrical 3D pattern around the antenna. Because the vibrational frequencies of the charge are extensive, there is a large range of possible frequencies and wavelengths of e&m waves, ranging from radio waves to gamma rays. The electromagnetic wave is the energy that moves away from the vibrating charges at the speed of light, $c = 3 \times 10^8$ m/s. If the electron wiggles faster, the frequency of the wave increases, but the speed at which the light moves away from the vibrating charge stays the same.

27.3.2 ELECTROMAGNETIC SPECTRA

The entire range of e&m waves at all possible frequencies is known as the *electromagnetic spectrum*.

Radio waves are what we call the electromagnetic waves with the longest wavelength and thus the lowest frequency. Radio waves range from as large as we can measure, kilometers in length, down to approximately 1 cm in length. Since these are the longest electromagnetic waves, they are produced by the largest groups of oscillating charges. Radio towers are approximately ¼ the wavelength of the waves they produce, and some of the natural sources of these waves are the convection loops of charged particles in stars, like our sun. In fact, the radio waves generated by the sun correspond to locations of sun spots where free-charge, called plasma, is circulating around the spots.

Microwaves range in length from approximately 30 cm to about 1 mm. They are produced by small oscillating or rotating groups of charges, such as in a microwave oven. In the microwave oven is a device called a magnetron, which is a vacuum cavity in which a magnet field guides a stream of electrons into a circular path. Since there is a centripetal acceleration, the electrons radiate at a frequency that matches the frequency of orbit of an electron. Remember that the frequency is the inverse of the period of oscillation, and the period can be found with simple centripetal force calculation. The microwaves in the microwave oven are rotating with a frequency that matches one of the rotational frequencies of the water molecule. Thus, these waves resonate with the water molecules in food, making the molecules spin faster and faster, thus heating up the food.

Infrared waves (IR) range in length from about 1 mm to 700 nm. Water molecules have strong absorptions in this part of the spectrum, so when the water molecules in our skin are exposed to these waves, we perceive it as heat. Also, as sunlight strikes the surface of the earth, it is absorbed, which in turn heats up the surface of the earth. The surface then radiates in the infrared. Molecules, such as water, carbon dioxide, methane, and other greenhouse gasses absorb the IR, delaying its escape out to space. Since the IR energy remains in the atmosphere longer, the planet reaches a higher equilibrium temperature. This is the Greenhouse Effect.

Visible light ranges from 700 nm to 400 nm. One explanation of why we see this part of the spectrum is because it is transparent to water. Thus, creatures that may have been evolving in water that could sense this part of the electromagnetic spectrum would have an advantage for survival. You may have noticed that you can see hundreds of feet down in clear water.

Ultraviolet radiation ranges from 400 nm to approximately 10 nm. These waves have wavelengths that are too short for us to see, but they are absorbed by our skin and may cause damage to our cells.

X-rays range in wavelength from 10 nm to 0.001 nm. These waves are short enough, so their frequency is high enough $\left(\nu = \frac{c}{\lambda}\right)$, so the photons have high-enough energy $\left(E_p = h\nu\right)$ to penetrate human tissue, but not bone. Thus, they are used extensively in medicine. The generation of X-rays will be covered in Chapter 29, since it is best understood using the quantum nature of light.

Gamma rays range from 0.001 nm to the limit of the smallest we can measure. The source of these rays is nuclear processes, which will be covered in Chapter 30.

A summary of these groups of e&m waves and their frequency and wavelength ranges is given in Table 27.1.

TABLE 27.1
Electromagnetic Waves Frequency and Wavelength Range

Radiation Name	Frequency	Wavelength
Radio waves	<300 MHz	>1 m
Microwaves	300–750,000 MHz	.4 mm–1 m
Infrared	750 GHz–430 THz	700 nm–.4 mm
Visible	430–750 THz	400–700 nm
Ultraviolet	750–6,000 THz	5–400 nm
X rays	6,000–50,000,000 THz	.006–5 nm
Gamma rays	>50,000,000 THz	<.006 nm

27.4 INTERFERENCE AND DIFFRACTION

27.4.1 YOUNG'S DOUBLE SLIT EXPERIMENT

In 1807, Thomas Young performed the first experiment that began to convince people that light is a wave. A modern version of this experiment, depicted in Figure 27.4, consists of a beam of monochromatic and coherent light directed onto a pair of parallel, thin slits cut into a screen.

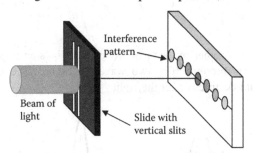

FIGURE 27.4 Young's double slit experiment.

A common way to produce the double slit for the experiment is by painting a microscope slide and etching out two thin, closely-spaced, parallel slits with a razor blade. The light shown upon the slits must be monochromatic light, which means it is made up of light that is a single color or wavelength. Most light is not exactly one and only one color. Even if a filter is used, it all has a spread in wavelength or frequency, which is called the band-width of the light. Coherent light is made up of light waves that are all in step, or in phase. Producing light with these special characteristics of monochromaticity and coherence was a challenge for Young, but for us it is as simple as turning on a laser. It turns out that laser light, simply by how it is produced, has a small bandwidth and is coherent enough to do the double slit experiment. The lasing process will be described as part of a synthesis opportunity in a few chapters. At first, you may have suspected that two thin bright lines of light would show up on a screen on the other side of the double slits, but that is not what happens. Instead, a pattern of symmetric bright spots separated by darker areas is produced, which is called an interference pattern. Young developed an explanation of the interference pattern using the properties of waves. Thus, this experiment and the subsequent explanation are considered one of the first experimental proofs that light exhibits wave properties.

27.4.2 DIFFRACTION

The explanation begins by considering what happens to any wave that interacts with a barrier. Consider a water wave in Figure 27.5 that is moving from left to right in the left section of the figure.

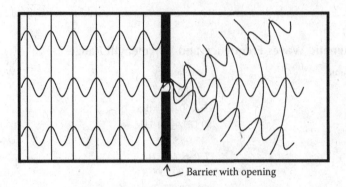

Barrier with opening

FIGURE 27.5   Diffraction of a wave that is moving from left to right.

The crests of the waves are signified with the lines, which are known as wavefronts. As the wave on the left to strikes a barrier, some of the wave will come through the opening in the middle of the barrier. When this happens, the waves fan out as described in Figure 27.5. This fanning out of the wavefronts is called **_diffraction_** and is a property of all waves, not just water waves. It is why you can hear what is happening in another room of a building even when you don't have a clear line of sight.

27.4.3   INTERFERENCE

As presented in Chapter 15, another property of waves that plays an important part in the double slit experiment is that they interfere. The two waves in Figure 27.6 demonstrate constructive interference on the left and destructive on the right.

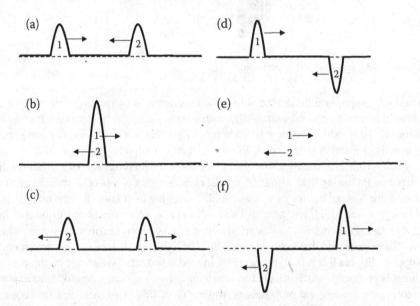

FIGURE 27.6   Interference of waves, constructive on the left and destructive on the right.

In part A, the two waves each have amplitudes in the same direction, up in this case, and are moving toward each other. In part B, the two waves constructively interfere at the instant of time that the waves overlap at the same position and produce a single peak with an amplitude equal to the sum of the amplitudes of the two waves, 1 and 2. After the interference in part C, the two waves continue to move off in the direction they were moving before the constructive interference. In part D, the two waves have two equal amplitudes in the opposite direction, 1 is up and 2 is down, and are moving toward each other. In part E, the two waves constructively interfere and produce a zero amplitude, flat spot. After the interference, the two waves in part F, the two waves continue to move off in the direction they were moving before the destructive interference occurred.

Given a system like the one described in Figure 27.7, in which there are two small openings an equal distance from the center of the system, waves will diffract from both openings. The waves from each opening will constructively interfere at the location in which the wavefronts overlap, like along the dashed lines in Figure 27.7. If a screen is located at the spots where the wavefronts overlap, there will be a maximum amplitude. If these are light waves, the larger the amplitude of the wave, the brighter the spot.

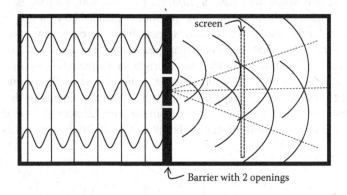

FIGURE 27.7   Arrangement with two small openings.

Referring to Figure 27.8, consider a maximum of the electric field produced at slits 1 and 2 at the same instant in time.

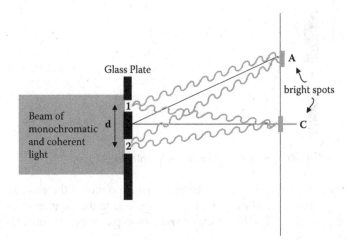

FIGURE 27.8   Double slit illuminated by monochromatic and coherent light beam and an example of the waves responsible for constructive interference producing the bright spots.

At each slit, the wavefronts fan out and diffract, and light is sent in all directions on the plane of the double slit experiment. Exaggerating the length of the waves to make it easier to visualize, but keeping all the waves the same length, due to the monochromaticity of the light, the relationships of the waves from each slit to form the interference pattern can be established. Starting with the central bright at location C, the distance from each of the slits to location C in the center of the screen is the same. Since the waves coming from each slit are in phase, because the light is also coherent, whatever phase the wave from slit 1 arrives with at C will be the same as the phase for the wave from slit 2. So, there should be a bright spot at the center point due to constructive interference.

Taking a closer look at Figure 27.8 reveals, there are 10 wavelengths from slit 1 to point C, and also from slit 2 to point C, to produce the central bright spot. So, the number of wavelengths from each slit to form the central bright spot is the same. So, there is a wavelength difference of zero, $\Delta\lambda = 0$. For bright spot A, there are 10.5 wavelengths from slit 1 to point A and 11.5 wavelengths. Therefore, this location of the first bright spot from the central bright, which is called the first order bright, results from a wavelength difference of one wavelength, $\Delta\lambda = 1$. The next bright spot further from the central bright will have a difference of two wavelengths, $\Delta\lambda = 2$, and so on. Therefore, if the variable of the order of the bright spot is established as $n = 0, 1, 2, 3, 4, \ldots$ this number corresponds to the number of the bright spots from the center and also the wavelength difference of the distance from the two slits to that spot. The pattern continues on for all the visible bright spot and is also symmetric about the central bright so the pattern is the same on either side. Taking a closer look at the bright spot produced at point A reveals the geometric relationship that can be used to find the wavelength of light from the measurements in the experiment. As shown in more detail in Figure 27.9, the distance from the point between the two slits to the screen is labeled L, and the distance from the central bright to the nth bright spot is labeled x.

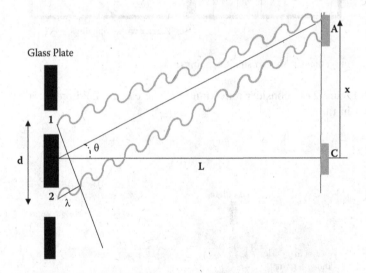

FIGURE 27.9   The wavelengths from the slits to the first order bright spot.

Thus, the tangent of the angle created by the line from the center of the slits to the central bright spot and the line from the center of the point between the slits to the location of the nth bright spot is labeled, θ, and is ratio of x to L. This geometric relationship of the experimental result is given in equation (27.4) as

$$\tan(\theta) = \frac{x}{L} \qquad (27.4)$$

Zooming in even further on the slits in Figure 27.9, a line from the center of slit 1 through a point on wave 2, is one wavelength, λ, from slit 2, which produces another triangle with the same angle theta (θ). Notice that the right angle between the line, L, and the slide is split by the line that defines the angle θ, by L and x. So, the angle ϕ must be 90°-θ. Thus, the triangle in Figure 27.10 with opposite side of λ and hypotenuse of d is related to the sine of theta (θ), which is the ratio of the wavelength of the light to the distance between the slits, d. This can be expressed as $\sin(\theta) = \frac{\lambda}{d}$.

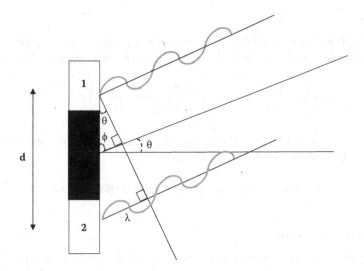

FIGURE 27.10 Zooming into the triangles at the location of the double slits.

If the second order bright is measured instead of the first, as in the diagrams, the sine would be related to twice the wavelength, since the difference in the distance traveled by light from the two slits increases by integer multiples of wavelength so that constructive interference can be achieved.

Thus, for the nth order bright spot, the relationship is normally rearranged into the form of equation (27.5) as

$$n\lambda = d\sin(\theta). \tag{27.5}$$

This expression and equation (27.4) provide the expressions needed to solve most problems associated with the bright spots in the double slit experiment. It is important to point out that the two lines from each slit to the bright spots are not perfectly parallel, so these expressions approximate the situation. This is normally a good approximation, since the slits are commonly on the order millimeters apart and the screen is on the order a meter away, so the lines are very close to parallel.

Between the bright spots, the light gets dimmer to a point of minimum brightness, which is known as a *dark spot*. These dark spots are at locations of destructive interference that occur at path length differences of one-half of a wavelength. Therefore, the expression to find the location of the dark spots looks similar to the one for bright spots, except for the addition of the ½. This expression is given in equation (27.6) as

$$\left(n + \frac{1}{2}\right)\lambda = d\sin(\theta). \tag{27.6}$$

where $n = 0, 1, 2, \ldots$ and the 0th dark spot is between the central and the first order bright and so on.

Double slit example

Monochromatic, coherent light with a wavelength of 560 nm is incident on two parallel narrow slits separated by 0.2 mm, and a diffraction pattern is produced on a wall 1.5 m away. Find the distance from the central bright on the wall to the second order bright spot on the wall.

Solution

Use equation (27.5) to find the angle, θ, to the second order bright spot, with the second order so $n = 2$, the slit separation is $d = 0.2 \times 10^{-3}$ m, and the wavelength gives $\lambda = 560 \times 10^{-9}$ nm.

$$\theta = \sin^{-1}\left(\frac{n\,\lambda}{d}\right) = \sin^{-1}\left(\frac{(2)(560 \times 10^{-9}\,\text{m})}{0.2 \times 10^{-3}\,\text{m}}\right) = 0.321°$$

Then, use equation (27.4) to find the distance from the central bright to the second order bright on the screen that is $L = 1.5$ m away, as seen in Figure 27.9.

$$x = (L)\,\tan(\theta) = (1.5\,\text{m})\,\tan(0.321°) = 0.0084\ \text{m} = 8.4\ \text{mm}$$

27.4.4 DIFFRACTION GRATING

A diffraction grating is a critical part of many spectroscopic devices that scientists use to measure the wavelength of light generated by samples that are under investigation. The diffraction grating consists of hundreds to thousands of equally spaced slits, side by side, per millimeter. As depicted in Figure 27.11, a diffraction grating is made up of a piece of glass with thousands of parallel and evenly spaced scratches.

FIGURE 27.11 Diffraction grating.

Unlike the depiction in Figure 27.11, the scratches are normally too small to see, and if you looked at a diffraction grating all you would see is a "rainbow" of color. Conveniently, the slit spacing is the center-to-center distance between adjacent slits. The angular positions of the maxima and minima are the same as for the double slits, but the maxima are much narrower and the minima are much wider. The maxima appear as thin vertical lines and are referred to as lines rather than fringes. The manufacturer typically provides the number of scratches or lines per unit of distance rather than the slit spacing, for instance, the number of lines per mm. The slit spacing can be thought of as the number of millimeters per line, so the slit spacing is simply the reciprocal of the number of lines per unit of distance.

As the light from each slit diffracts and interferes with light from another slit, the pattern produced is explained by the same processes as in the double slit, with the center-to-center distance between each adjacent slit as the d in equation (27.5).

Diffraction grating example

Find the wavelength of the light that is shown upon a diffraction grating with 1200 line per millimeter if the first order bright spot is measured at an angle of θ = 39.675°, as shown in Figure 27.12.

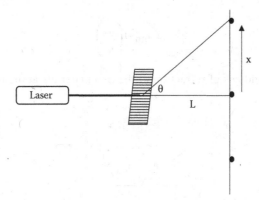

FIGURE 27.12   Diffraction grating example.

Solution

Since there are 1,200 lines per millimeter, the slit spacing is:

$$d = \frac{1}{1200} \text{ mm} = 8.3333 \times 10^{-7} \text{ m}$$

Employing equation (27.5)

$$\lambda = \frac{d \ \sin(\theta)}{n} = \frac{8.3333 \times 10^{-7} \text{ m} \ \sin(39.675°)}{1} = 5.32 \times 10^{-7} \text{ m} = 532 \text{ nm}$$

This is green light.

27.5   POLARIZATION

Polarization is another property of light, which is best explained with a wave model. As described in the beginning of the chapter, a vibrating charge generates an electromagnetic wave. If the charge is restricted to vibrate in one orientation, as depicted in Figure 27.1, the electromagnetic wave produced will oscillate in one orientation. The orientation of the electric field of an electromagnetic wave is the wave's polarization. Thus, the electromagnetic wave in Figure 27.1 is vertically polarization. Since the vibrating electrons in most light sources, including the sun and light bulbs, can vibrate in any direction the light they produce has no polarization. Thus, most light has a randomly changing polarization, which is called unpolarized or randomly polarized light. The symbol for unpolarized light is given in Figure 27.13.

FIGURE 27.13   Symbol for unpolarized light.

Unpolarized light can be polarized by both reflection and absorption. As unpolarized light waves reflect off surfaces at certain angles, only the components of the electric field that are parallel to the surface reflect. First explained by David Brewster in the early 1800s, the angle at which this phenomenon occurs is called Brewster's Angle, θ_B, and is given by equation (27.7) as

$$\theta_B = \tan^{-1}\left(\frac{n_2}{n_1}\right). \tag{27.7}$$

where n_1 and n_2 are the indices of refraction of the two materials at the reflection interface (see Figure 27.14).

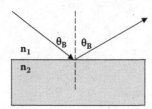

FIGURE 27.14 Brewster's angle example.

For example, if light is reflecting off the calm surface of the lake, the angle at which all the reflected light will all have the same polarization parallel to the surface of the water is:

$$\theta_B = \tan^{-1}\left(\frac{1.33}{1}\right) = 53.06°$$

Remember that the angle of incidence and the angle of refraction are measured from the normal line, so the angle of this polarized light is approximately $(90° - 53°) = 37°$ above the horizontal. Since the reflected light waves all have the same orientation, they add together to form a bright light we call *glare*. Polarized sunglasses are designed to absorb this horizontally oriented glare.

Unpolarized light can also be polarized by absorption. Some naturally forming crystals, such as calcite ($CaCO_3$), have a structure such that it has different indices of refraction along different planes of the crystal. When looking at an object through a crystal such as calcite, you will see two images that are polarized in two different orientations. Today we have employed this distinct property of birefringence in many applications, such as in lasers and liquid crystal displays.

In addition to the naturally-forming crystals, there are devices we have developed that will absorb light with a preferred polarization. These polarizers absorb light along long molecular chains, commonly in a polymer. These molecular chains are oriented all in the same direction in the formation process of the plastic by exposing the molten plastic to a strong electric field. As the plastic cools, the molecular chains remain in the same orientation.

A common application of these absorption polarizers are the lenses in polarized sunglasses. Since most of the surfaces from which glare is produced are horizontal, the polarizers in polarized sunglasses are oriented to absorb light that is horizontally polarized. The rest of the light around us is commonly unpolarized, so on average, 50% of it passes through the lenses, so we can still see even when the glare is cut out.

As shown in Figure 27.15, if the intensity of the incoming light is I_0, then the intensity of the light that gets through is called I_1. The expression for the intensity of the randomly polarized light passing through a polarizer is given by equation (27.8) as

$$I_1 = \frac{1}{2}I_0. \tag{27.8}$$

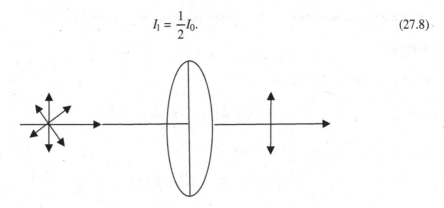

FIGURE 27.15 Half (50%) of unpolarized light passes through a polarizer, and it is polarized with the orientation of the polarizer.

For polarized light incident on a polarizer, the intensity of the light transmitted through the polarizer is a function of the angle, θ, between the orientations of polarizer and the polarization of the incident light, as described in Figure 27.16.

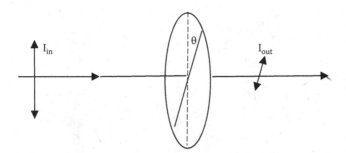

FIGURE 27.16 A polarizer oriented at an angle different than the orientation of light wave.

Malus's Law is the mathematical relationship of the outgoing intensity (I_{out}) to the incoming intensity (I_{in}) of light for a polarizer, which is given in equation (27.9) as

$$I_{out} = I_{in} \cos^2(\theta). \tag{27.9}$$

Example of polarization

A beam of unpolarized light, with an intensity of $8 \frac{W}{m^2}$, is incident on a polarizer with a vertical orientation. The light exiting the vertical polarizer is incident on a polarizer at an angle of $\theta = 20°$ to the vertical, as depicted in Figure 27.17. What is the intensity of the light after is passes through the second polarizer?

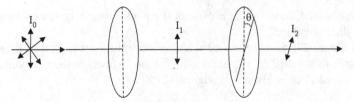

FIGURE 27.17 Arrangement for the polarizer example with two polarizers.

Solution

First, use equation (27.8) to find I_1:

$$I_1 = \frac{1}{2}I_o = \frac{1}{2}\left(8 \frac{W}{m^2}\right) = 4 \frac{W}{m^2}$$

Second, apply equation (27.9) to find I_2:

$$I_2 = I_1 \cos^2(\theta) = 4\frac{W}{m^2}\cos^2(20°) = 3.53\frac{W}{m^2}$$

27.6 ANSWER TO CHAPTER QUESTION

X-ray crystallography is the process of illuminating a sample with X-rays and measuring those that are scattered. Depicted in Figure 27.18, this process begins with X-rays directed at a crystal, and the intensity of the scattered X-rays is measured at a series of angles, labeled $2(\theta)$ in the diagram. The angles at which the maximum signals are measured correspond to the angles at which constructive interference occurs as X-rays scatter off layers of the atoms.

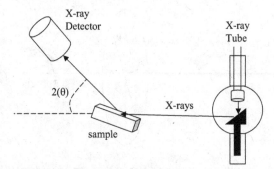

FIGURE 27.18 Arrangement of an X-ray scattering experiment.

Using the geometry displayed in Figure 27.19, notice that the waves indecent upon the crystal and reflected off the crystal are in phase.

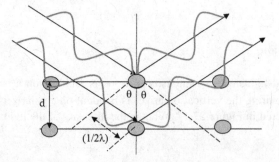

FIGURE 27.19 The scattering of X-rays off layers of atoms.

In Figure 27.19, the triangle with the hypotenuse, d, and the angle, θ, must have an opposite side with a length of (1/2λ) to keep the waves in phase. The geometry of this arrangement can be expressed in equation (27.10) as

$$n\lambda = 2d \, \sin(\theta). \tag{27.10}$$

This expression is similar to equation (27.5) but with an extra factor of 2 because the X-rays reflect off the layers of the atoms instead of passing through the double slits. With knowledge of the wavelength (λ) and measurement of the angle (θ), the distance (d) between the planes of the atoms in the crystal can be computed. By rotating the crystal through all possible orientations, measurements of the distance between the planes of the atoms can be found, and a picture of the atomic structure can be developed. This process reveals the structure of the atoms in the crystal and how we know that the sodium and chlorine atoms in salt are in a cubic structure.

X-ray crystallography is also the experimental process that Rosalind Franklin employed in the early 1950s with DNA to generate an X-ray diffraction pattern similar to the one sketched in Figure 27.20. Watson and Crick then interpreted the pattern as generated by the scattering of X-rays off a double helix, and the rest is history.

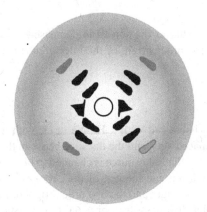

FIGURE 27.20   X-ray diffraction pattern from the scattering off DNA.

27.7   QUESTIONS AND PROBLEMS

27.7.1   MULTIPLE CHOICE QUESTIONS

1. Which is faster in a vacuum, visible light or X-rays?
   A. visible light
   B. X-rays
   C. neither – both travel at the same speed in a vacuum
2. Some lasers produce light in a region outside the visible spectrum, and then a crystal (called a doubler) is used to double the frequency so that the output of the doubler is in the visible part of the spectrum. To which part of the electromagnetic spectrum does the output of the laser belong if the wavelength of the output of the doubler is 532 nm?
   A. radio waves
   B. microwaves
   C. infrared
   D. visible
   E. ultraviolet
   F. X-rays
   G. gamma rays

3. What happens to the intensity of light if you double the magnitude of the maximum value achieved by the electric field in its oscillations?
   A. The intensity becomes one-fourth what it was.
   B. The intensity becomes one-half what it was.
   C. The intensity remains what it was.
   D. The intensity becomes twice what it was.
   E. The intensity becomes four times what it was.

4. Red laser light is incident normally on two, narrow, closely-spaced slits a distance away from a screen, on which a diffraction pattern is produced. What happens to the spacing between the fringes of the diffraction pattern if the light is changed from red to blue?
   A. The spacing increases.
   B. The spacing decreases.
   C. No change occurs.

5. At the locations of minimum brightness, darks, in a diffraction pattern the light waves meeting at the screen, have a phase difference of:
   A. $0°$
   B. $90°$
   C. $180°$
   D. $270°$

6. If the distance from the slits to the screen is decreased, will the spacing between the central bright and the first order bright of the interference pattern increase, decrease, or remain the same?
   A. increase
   B. decrease
   C. remain the same

7. An interference pattern is created with a He-Ne laser, with a wavelength of 632.8 nm, shown on two closely spaced slits, and the distance between the central bright and the first order bright is measured. If the light source is changed to blue light from an argon ion laser with a wavelength of 488 nm and the rest of the experiment is left unchanged, will the spacing between the central bright and the first order bright of the interference pattern with the argon laser be greater than, less than, or equal to the distance for the HeNe laser?
   A. greater than
   B. less than
   C. equal to

8. If an interference pattern is created with a He-Ne laser with a wavelength of 632.8 nm shown on two closely spaced slits, and then the distance from the slits to the screen is decreased, will the angle (θ) between the line from the center of the slits to the central bright and the line from the center of the slits to the first order bright spot of the interference pattern increase, decrease, or remain the same?
   A. increase
   B. decrease
   C. remain the same

9. For light waves, what is the polarization direction of the light?
   A. the angle between electric and magnetic fields of the wave
   B. the change in the speed of the wave as it enters another medium, besides a vacuum
   C. the orientation of the wave's electric field

10. For a polarizer that is a flat plastic sheet of material, what is the polarization direction of the polarizer?
    A. the polarization direction of the light that it lets through
    B. the angle that the plane of the polarizer makes with the vertical
    C. the angle between the plane of the polarizer and the direction of propagation of the light

27.7.2 PROBLEMS

1. Two slits, 4 μm apart, are illuminated by light from a laser that produces an interference pattern (interferogram) on a screen 0.5 m away from the slits. The third order bright of this pattern is 21.7 cm away from the center of the central bright. Find the angle to the second order bright of this interferogram.

2. During an experiment done in a lab, a laser is shown through a pair of slits, separated by 0.2 mm, and a diffraction pattern is seen on a wall 5.0 m. The seventh order bright of the diffraction pattern on the wall is located 9.45 cm away from the central order bright. For this experiment:
   a. Sketch and label the experimental set-up.
   b. Sketch the diffraction pattern and label the central and seventh order bright.
   c. Calculate the wavelength of the laser in nm.

3. What is the wavelength (nm) of monochromatic light incident on two slits 0.015 mm apart, which produces the fifth order fringe at an angle of 8.0°?

4. What is the wavelength (nm) and frequency (Hz) of monochromatic light incident on two narrow slits, separated by 0.048 mm, which produces a diffraction pattern on a screen 5.00 m away, in which the first order fringe is 6.5 cm away from the central bright fringe?

5. Two narrow slits separated by 1.0 mm are illuminated by 544 nm light. Find the distance (cm) between the central bright and the third order bright fringes on a screen 3.0 m away from the slits.

6. Light from a He-Ne laser is monochromatic, coherent, and at a wavelength of 632.8 nm. In a double slit experiment, light from a He-Ne laser is incident normal on a pair of narrow, closely-spaced, parallel, and vertical slits. A screen located 2.5 m away from the slits is oriented so that it is parallel to the plane of the slits and perpendicular to the path of the light beam. The interference pattern produced is given in Figure 27.21. Note that the ellipse with the C above it is the central bright of the interference pattern. Compute the distance between the slits.

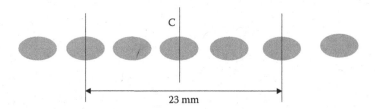

FIGURE 27.21   Diagram for problem 6.

7. A beam of monochromatic, coherent, cyan light, with a wavelength of 500 nm, is incident on two thin slits spaced 16.67 μm apart. An interference pattern is seen on a screen 1 m from the double slits. Compute the distance between the central bright spot and the fourth order bright spot on the screen.

8. A horizontal beam of completely unpolarized light of intensity 100 W/m² is normally incident on a polarizer whose polarization direction makes an angle of 30° clockwise (as viewed from behind the polarizer, where we define the light source to be behind the polarizer) from the vertical. The light that passes through that polarizer is normally incident on a polarizer whose polarization direction makes an angle of 30° *counter-clockwise* (as viewed from behind the polarizer, where we define the light source to be behind the polarizer). What is the intensity of the light that makes it through the second polarizer?

9. A vertical beam of polarized light of intensity 68.0 W/m^2 is shining straight downward on the horizontal surface of a flat polarizer. The polarization direction of the light is 10.0° west of south. The polarization direction of the polarizer is eastward. Find the intensity of the light that makes it through the polarizer.

10. X-rays with a wavelength of 0.0567 nm are scattered off a crystalized molecule, and the minimum angle of constructive scattering is 11°. What is the spacing between the atomic layers in the crystalized molecule?

28 Quantum Nature of Matter

28.1 INTRODUCTION

This chapter is the first of three chapters in this volume that are focused on atomic and nuclear physics. In an effort to connect this chapter to the previous discussion of optics and provide some historical context, the line spectra generated by energized gases is the starting point of the discussion about the quantized nature of matter. The Bohr model of the hydrogen atom provides the structure to described the origin of this discrete spectrum generated by transitions of electrons in the atom. This model of the hydrogen atom also provides an opportunity to review many of the concepts of energy, momentum, and angular momentum presented in the beginning of this volume. Next, the concept of De Broglie's matter waves is introduced as a way to explain the assumptions that Bohr made in developing his model of the hydrogen atom. This chapter concludes with a description of quantum mechanics and some of the fascinating conclusions about the structure of matter that can be explained with this fundamental theory.

28.2 CHAPTER QUESTION

In most chemistry books there are images of orbitals, similar to those sketched in Figure 28.1. The question is are these orbitals? This question will be answered at the end of this chapter in the presentation about quantum mechanics.

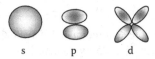

s   p   d

FIGURE 28.1 Orbitals of an atom.

28.3 QUANTIZATION OF MATTER

28.3.1 SPECTRA

The starting point of an investigation of the structure of matter starts with an analysis of the light generated when different substances are heated. When a solid or a liquid is heated to a high temperature, it begins to glow at a visible wavelength. First, it glows red, then yellow. Then, if the object is hot enough, it will glow across the entire visible spectrum, so it looks white. If the light from the glowing object is passed through a prism, the visible spectrum is seen, as depicted in Figure 25.2. As described in Chapter 25, this result is explained as the black body radiation from the vibrating charged objects (electrons) in the solid or liquid. This theory of vibrating sources of light was developed by Max Planck and began the development of the quantum theory of matter. Planck's theory was able to explain that as more energy is pumped into the system, the kinetic energy of the oscillators increases, so they vibrate quicker, producing electromagnetic waves with higher frequencies, and thus shorter wavelengths. This process moved the color produced from red, to orange, to yellow, and so on.

When a gas, like hydrogen at a low pressure contained in a glass tube, is put across a high voltage, it also begins to glow. But in the case of a gas, if the voltage is increased the light gets brighter, but the color does not change. In fact, when the light from an energized gas is passed through a prism, as demonstrated in Figure 28.2, a line spectrum is produced.

DOI: 10.1201/9781003308072-29

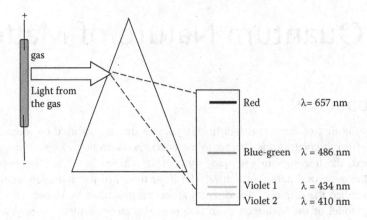

FIGURE 28.2   Line spectrum of a hydrogen gas.

The analysis begins with hydrogen, since it is the simplest gas comprised of one proton and one electron. When hydrogen gas is energized, it glows a magenta color, because its line spectrum has strong lines in the red and blue-green, so together we see the light as magenta. When the light from the hydrogen gas is shown through a prism, the line spectrum is revealed. As depicted in Figure 28.2, the line spectrum consists of a red line at 657 nm, a blue-green line at 486 nm, and two violet lines at 434 nm and 410 nm. These violet lines are dim, so they have only a slight effect on the color we see when we look at an energized hydrogen gas tube.

28.3.2   BALMER SERIES

The first mathematical expression able to predict the wavelengths of light in the line spectrum generated by hydrogen was produced in 1885 by Johann Balmer. Although Balmer published the relationship in a somewhat different form, the modern equivalent of the Balmer formula is given in equation (28.1) as

$$\frac{1}{\lambda} = R\left(\frac{1}{n^2} - \frac{1}{m^2}\right). \tag{28.1}$$

with m and n being positive integers and $n < m$. The constant R, known as the Rydberg constant, was adjusted to make the formula produce the wavelengths observed and has a value of $R = 1.097 \times 10^7 \text{ m}^{-1}$. The Balmer series gives no physical explanation for the lines, just an equation that predicts the wavelength of the colors emitted by hydrogen.

Example

What is the value of the wavelength for values of $n = 2$ and $m = 3$?

Solution

Plugging in $n = 2$, $m = 3$, and $R_H = 1.097 \times 10^7 \text{ m}^{-1}$ into equation (28.1) gives:

$$\frac{1}{\lambda} = R\left(\frac{1}{n^2} - \frac{1}{m^2}\right) = 1.098 \times 10^7 \text{ m}^{-1}\left(\frac{1}{2^2} - \frac{1}{3^2}\right) = 0.15236 \text{ m}^{-1}$$

Next, take the reciprocal of the previous step:

$$\lambda = 6.57 \times 10^{-7}\, m = 657 \times 10^{-9}\, m = 657\, nm.$$

This is the wavelength of the red spectral line of hydrogen.

The set of wavelengths corresponding to $n = 2$ and $m = 3, 4, 5,$ and 6 is called the *Balmer series,* and the wavelengths produced in this series are in the visible part of the e&m spectrum. The set of wavelengths corresponding to $n = 1$ and $m = 2, 3, 4, 5, \ldots$ is called the *Lyman series,* and wavelengths produced in this series are in the ultraviolet part of the e&m spectrum. The set of wavelengths corresponding to $n = 3$ and $m = 4, 5, 6, \ldots$ is called the *Paschen series,* and wavelengths produced in this series are in the infrared part of the e&m spectrum. The reason the wavelengths conform to this mathematical formula, with the experimentally determined value of R, was successfully explained by a model of the hydrogen atom published by Niels Bohr in 1913 and now known as the Bohr model.

28.3.3 BOHR'S MODEL

In 1913, Niels Bohr was able to derive the Balmer formula by making a series of revolutionary postulates. The Bohr Theory of the hydrogen atom gained acceptance from this success. Bohr's model is based upon the assumption that the electrons in an atom orbit around the nucleus, which is made up of an integer (Z) number of protons. The analysis begins with a classical study of forces and energy of an orbiting electron around a nucleus.

Since the electron is orbiting around the nucleus, the centripetal force is equal to the Coulomb force (F_c) of electrostatic attraction between the electron and the nucleus (see Figure 28.3):

$$\frac{mv^2}{r} = k\,\frac{q\,Zq}{r^2}$$

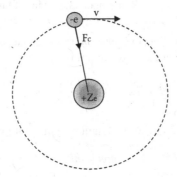

FIGURE 28.3 Bohr model of the hydrogen atom.

Since the charge of the electron and the proton is the same magnitude, the variable used to indicate the elementary electrical charge, an (e), is used for the charge instead of (q). If both sides are multiplied by r and both sides are multiplied by ½, an expression for the kinetic energy of the orbiting electron is derived

$$K = \frac{1}{2}\,mv^2 = k\,\frac{Ze^2}{2r}$$

The potential energy of the electron is $U = F\,r = -k\dfrac{Ze^2}{r}$

So, the total energy of the electron in the orbit, is the sum of the kinetic and the potential energy of the electron in the orbit,

$$E = K + U = k \frac{Ze^2}{2r} + -k \frac{Ze^2}{r}.$$

Since both terms are similar with only a $-\frac{1}{2}$ difference, the sum is just,

$$E = -k \frac{Ze^2}{2r} = -K$$

and the total energy is just the negative of the kinetic energy. The negative total energy implies that the electron is bound to the proton because the potential energy is a greater negative value than the positive kinetic energy.

If the velocity of the orbiting electron was known, the kinetic energy could be found, and thus the energy of the orbits could be found. If the radius of the orbit was known, the potential energy could be found, and thus the energy of the orbits could be found. Unfortunately, since neither the velocity (v) of the electron or the radius (r) of its orbit was known, the analysis could not be taken any further until an assumption was made.

The first key assumption, named Postulate 1, is that the electrons can only exist in special orbits in which the angular momentum $\left(\overrightarrow{L} \right)$ of the orbits is quantized. This quantization is formalized in equation (28.2) as

$$L_n = n\hbar. \tag{28.2}$$

where $\hbar$, which is known as "h-bar", is defined as $\hbar = \frac{h}{2\pi}$, with Planck's Constant ($h = 6.63 \times 10^{-34}$ Js), the value of $\hbar = 1.0546 \times 10^{-34}$ J·s. Remember that Max Planck's constant was introduced to explain the shape of the black body curve. Angular momentum was defined in Chapter 18 as equation 18.2 as $\overrightarrow{L} = I\ \overrightarrow{\omega}$. For an object of mass, m, moving in a circular path with a radius, r, and a velocity, v, the magnitude of the angular momentum is:

$$L = I\omega = (mr^2)\left(\frac{v}{r}\right) = mvr.$$

Applying Postulate 1, equation (28.2) to the angular momentum of the orbiting electron gives:

$$mvr = n\hbar$$

Solving this expression for mv results in $mv = \frac{n\hbar}{r}$.

Both sides of this expression can be squared, then both sides multiplied by ½ and divided by the mass, m, of the electron to give an expression for the kinetic energy of the orbiting electron:

$$\frac{1}{2} \frac{(mv)^2}{m} = \frac{\left(\frac{n\hbar}{r}\right)^2}{2m}$$

$$K = \frac{1}{2}mv^2 = \frac{n^2h^2}{(2\pi)^2 r^2 2m} = \frac{n^2h^2}{8m\pi^2 r^2}$$

Combining the energy of the classical electron orbit with the quantization of angular momentum, the Bohr approach yields expressions for the electron orbit radii and energies.

Since the kinetic energy is the same magnitude as the total energy and opposite in sign, the total energy of the electron is:

$$E_n = -K = -\frac{n^2 h^2}{8 m_e \pi^2 r^2} = -k \frac{Z e^2}{2r}$$

Remember that a negative energy is acceptable, in the way a marble "orbiting" about the center of bowl, shown in Figure 28.4, has a negative energy if the top of the bowl is set to the zero height.

FIGURE 28.4 Marble orbiting around a bowl.

That is, the negative energy indicates that the electron is bound to the proton. The only variable in the last two terms on the right is r; therefore, an expression for r can be found by setting the last two terms on the right of the equation above.

$$-\frac{n^2 h^2}{8 m_e \pi^2 r^2} = -k \frac{Z e^2}{2r}$$

This gives the following expression for r_n in equation (28.3) as

$$r_n = \frac{\hbar^2}{Z m_e k e^2} n^2. \tag{28.3}$$

If all the quantities are plugged into the expression, the resulting constant is evaluated to 0.52918×10^{-10} m, which is known as the Bohr radius. One of the reasons that Bohr's model became accepted is that this value corresponds to the approximate size of the atom that other scientists were finding with scattering experiments. The expression for the radius r_n of the nth orbit is given in equation (28.4) as

$$r_n = (0.529 \times 10^{-10} \ m)\frac{n^2}{Z}. \tag{28.4}$$

Notice that not only do the orbits get bigger with increasing values of n, but the spacing between adjacent orbits increases because n is squared. Also, if there are more protons in the nucleus, the radii of the orbits decrease. This agrees with a greater pull on the orbiting electrons due to more protons in the nucleus.

Plugging equation (28.3) for r into the expression for energies of the orbiting electron gives

$$E_n = k \frac{Z e^2}{2\left(\frac{n^2 \hbar^2}{m_e k Z e^2}\right)}$$

Which can be simplified to equation (28.5) as

$$E_n = -\frac{Z^2 m_e k^2 e^4}{2 \hbar^2} \frac{1}{n^2} \tag{28.5}$$

Plugging in the values for the constants for hydrogen, which are $Z = 1$ since there is only one proton in the nucleus of hydrogen, the mass of an electron ($m_e = 9.109 \times 10^{-31}$ kg), the Coulomb constant $\left(k = 9 \times 10^9 \frac{Nm^2}{C^2} \right)$, the charge of an electron ($e = 1.602 \times 10^{-19}$ C), and h-bar ($\hbar = 1.0546 \times 10^{-34}$ $J \cdot s$), into equation (28.5) gives equation (28.6) as

$$E_n = (-2.1847 \times 10^{-18} \, J)\frac{1}{n^2}. \tag{28.6}$$

This value in Joules is normally converted into the unit of electron Volts (eV). This is the unit that is set to the amount of energy it takes to move an electron, with a charge of 1.602×10^{-19} C, across 1 eV. So, 1 eV = 1.602×10^{-19} J, then $E_1 = -13.6$ eV.

So, E_n is normally expressed as shown in equation (28.7) as

$$E_n = -13.6 \text{ eV } \frac{1}{n^2}. \tag{28.7}$$

The value -13.6 eV is referred to as the ground state energy of the hydrogen atom, since it is the lowest possible energy. It is analogous to the lowest orbit that the marble can have in the bowl. This is the quantization of the orbits. Unlike the classical example of the orbiting marble, which can have any speed and height below the rim on the top of the bowl, the electrons orbiting in the hydrogen atom have only specific set values of energy in which they can orbit.

An energy level diagram, with the first five orbits of electron in the hydrogen atom labeled, is given in Figure 28.5.

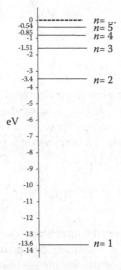

FIGURE 28.5   Energy level diagram of the electron orbits in the hydrogen atom.

Notice that the energy levels approach the zero, denoted with the dashed line in Figure 28.5, and the spacing between the levels gets smaller as the number of the level increases. This is explained by the result of equation (28.3), in which the value of -13.6 eV is divided by an ever-increasing number, which is the square of the number that specifies the level, n.

The *ionization energy* of an atom is the energy it would take to remove an electron from the ground state from an atom. In the case of hydrogen, this energy is 13.6 eV, but as the number of protons in the nucleus, Z, increases, the ionization energy increases with Z^2, as 13.6 eV (Z^2). So, in the marble analogy, if a marble has a kinetic energy greater than the potential energy of the edge of the bowl, the marble leaves the bowl.

To find the orbital frequency f_e of an electron in the hydrogen atom, the expression of the energy levels, and the relationship of the total energy and the kinetic energy derived in the previous section is employed.

$$K = -E_n = -\left(-\frac{Z^2 m_e k^2 e^4}{2\,\hbar^2}\right) = \frac{1}{2}m_e v^2$$

Set the velocity equal to the distance around a circular orbit, the circumference, divided by the time for one orbit, the period (T).

$$\left(\frac{Z^2 m_e k^2 e^4}{2\,\hbar^2}\right) = \frac{1}{2}m_e\left(\frac{2\pi r}{T}\right)^2$$

The period and the frequency are reciprocals, so $f_e = \frac{1}{T}$, the previous equation becomes

$$\left(\frac{Z^2 m_e k^2 e^4}{2\,\hbar^2}\right) = \frac{1}{2}m_e\,(2\pi r)^2 \left(f_e\right)^2$$

Solving for the frequency f_e gives

$$f_e = \sqrt{\frac{Z^2 k^2 e^4}{4\pi^2 \hbar^2 r^2}}$$

It is interesting to note that if the electron was orbiting the proton in its ground state orbit $(Z = 1)$, the frequency of the orbit would be $f_e = 6.59 \times 10^{15}$ Hz. If the electron was orbiting at this frequency, it should be radiating at this frequency, which corresponds to an e&m wave with a wavelength of $\lambda_e = 4.55 \times 10^{-8}$ m $= 45.5$ nm, which is in the vacuum ultraviolet part of the spectrum. All atoms do not constantly radiate away energy in this form, so Bohr needed to extend Postulate 1. Not only were the orbits quantized, but the electrons in these quantized orbits did not radiate like any other electron in an orbit.

Postulate 1

Electrons in atoms exist in specific quantized orbits around the nucleus, and the electrons in these quantized orbits do not radiate.

Bohr took his structure of the energy levels of the hydrogen atom and built the concept of how the spectral lines are produced with a second postulate.

Postulate 2

Electrons emit radiation (light) when they transition from a higher energy level to a lower energy level in the same atom. The energy of the photon (E_p) released in the process is the absolute value of the difference in the energy levels $|(E_f - E_i)|$, which is described in equation (28.8) as

$$E_p = \left|E_f - E_i\right| \tag{28.8}$$

28.3.4 Bohr Explains the Balmer Series

Assume electrons give up energy in the form of particles of light, photons, with energy equal to the energy difference in the jump, for hydrogen $(Z = 1)$.

Start with equation (28.8)

$$E_p = |E_f - E_i|$$

Then plug in the energy of a photon, equation (25.4), $E_p = h\nu$ and the energy of each level from equation (28.6), as $E_n = (-2.1847 \times 10^{-18} J)\frac{1}{n^2}$ to get

$$h\nu = \left|(-2.1847 \times 10^{-18} J)\frac{1}{n_f^2} - (-2.1847 \times 10^{-18} J)\frac{1}{n_i^2}\right|$$

which can be simplified to

$$h\nu = (2.1847 \times 10^{-18} J)\left[\frac{1}{n_f^2} - \frac{1}{n_i^2}\right]$$

From equation (25.3) $c = \lambda\nu$ the expression becomes

$$\frac{hc}{\lambda} = (2.1847 \times 10^{-18} J)\left[\frac{1}{n_f^2} - \frac{1}{n_i^2}\right]$$

Rearranging the expression results in

$$\frac{1}{\lambda} = \left(\frac{2.1847 \times 10^{-18} J}{hc}\right)\left[\frac{1}{n_f^2} - \frac{1}{n_i^2}\right]$$

Plugging in the values for h and c results in

$$\frac{1}{\lambda} = \left(\frac{2.1847 \times 10^{-18} J}{(6.63 \times 10^{-34} Js)\left(3 \times 10^8 \frac{m}{s}\right)}\right)\left[\frac{1}{n_f^2} - \frac{1}{n_i^2}\right]$$

which is

$$\frac{1}{\lambda} = (1.098 \times 10^7 \text{ m}^{-1})\left[\frac{1}{n_f^2} - \frac{1}{n_i^2}\right]$$

where $R_H = 1.098 \times 10^7$ m^{-1} is approximately the Rydberg constant and λ is the wavelength of the light produced in meters. If the second orbit is set to the final level of the transition and the initial levels are set to 3, 4, 5, and 6, the electron transitions that produce the photons are the same as those from the Balmer series. Therefore, the Bohr Model can be used to derive the Balmer series for the hydrogen atom.

Bohr model example

Calculate the wavelength of light that is generated in the transition of the electron from the $n = 4$ to the $n = 2$ state of the hydrogen atom.

Solution

Generate a sketch of the energy level diagram, as shown in Figure 28.6.

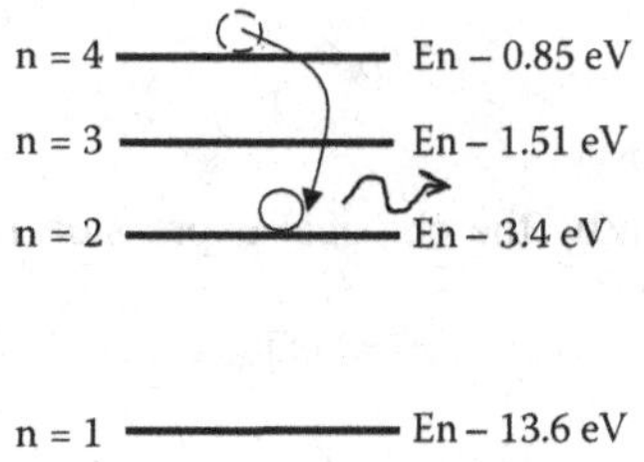

FIGURE 28.6 Energy level diagram for the Bohr model example.

Start the calculations with equation (28.7) to find the energy of the fourth and second levels:

$$E_i = -13.6 \text{ eV } \frac{1}{n_i^2} = -13.6 \text{ eV } \frac{1}{4^2} = -0.85 \text{ eV}$$

$$E_f = -13.6 \text{ eV} \frac{1}{n_f^2} = -13.6 \text{ eV} \frac{1}{2^2} = -3.4 \text{ eV}$$

Then, apply equation (28.8) to find the energy of the photon:

$$E_p = |E_f - E_i| = |-3.4 \text{ eV} - -0.85 \text{ eV}| = 2.55 \text{ eV}$$

Convert the photon energy to Joules:

$$E_p = 2.55 \text{ eV} \left(\frac{1.602 \times 10^{-19} J}{1 \text{ eV}} \right) = 4.09 \times 10^{-19} J$$

Find the frequency of the photon with equation (25.4), by solving for v.

$$v = \frac{E_p}{h} = \frac{4.09 \times 10^{-19} J}{6.63 \times 10^{-34} Js} = 6.17 \times 10^{14} \frac{1}{s} = 6.17 \times 10^{14} \text{ Hz}$$

Thus, the wavelength of the light can be found with equation (25.3), by solving for the λ.

$$\lambda = \frac{c}{v} = \frac{3 \times 10^8 \frac{m}{s}}{6.17 \times 10^{14} \frac{1}{s}} = 4.87 \times 10^{-7} \text{ m} = 487 \times 10^{-9} \text{ m} = 489 \text{ nm}$$

Bohr's model of the atom introduced quantized energy states for the electrons in atoms. Bohr's model produces values that are in good agreement with experiments for atoms with only one electron. Thus, it can be used to predict transitions in Hydrogen, singly ionized Helium (He^+), and doubly ionized Lithium (Li^{2+}), but does not work well for more complicated atoms.

28.4 THE WAVE NATURE OF PARTICLES

Bohr did not have an explanation for the quantization of the angular momentum of the electron in the hydrogen atom. He only knew that by using this quantization, the predicted energy levels matched the observed spectrum. An explanation for the quantization was developed by Louis de Broglie, who hypothesized that any particle that has momentum has wave properties with a wavelength that depends on the momentum of the particle, as given in equation (28.9) as

$$\lambda_D = \frac{h}{p} \qquad (28.9)$$

where the momentum is still given by the classical expression from equation (17.2) as

$$\vec{p} = m\vec{v}, \qquad (17.2)$$

The wavelength computed in equation (28.9) is known as the de Broglie wavelength and should not be confused with the wavelength of the light produced when an electron makes a transition. The de Broglie wavelength for the case of an electron is the wavelength of the electron.

De Broglie's idea provides the reason for the quantization of the angular momentum of the electron in the atom. De Broglie hypothesized that the only allowable orbits in the hydrogen atom are those whose circumferences are integer multiples of the wavelength of the electron, because electrons with other wavelengths would be destroyed by destructive interference as they orbited the nucleus. The electron wavelengths for the third and fourth orbits of the electron in the hydrogen atom are depicted in Figure 28.7.

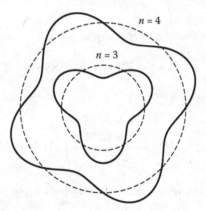

FIGURE 28.7   Depiction of the de Broglie waves for the $n = 3$ and $n = 4$ states of hydrogen.

Start at any point on the orbit and count the wavelengths around the circle. It should be clear that these are both integer number of wavelengths. The relation of these waves to the circumference of each orbit can be expressed in terms of the radius of the orbit as

$$2\pi r_n = n\lambda_n$$

Inserting $\lambda_n = \frac{h}{p} = \frac{h}{m_e v_n}$ in the above expression and remembering that the angular momentum of the orbit electron is $m_e v_n r_n$ is the angular momentum of the electron results in Bohr's quantization of the angular momentum, as in equation (28.2)

$$
\begin{aligned}
2\pi r_n &= n\frac{h}{m_e v_n} \\
m_e v_n r_n &= n\frac{h}{2\pi} \\
m_e v_n r_n &= n\hbar \\
L_n &= n\hbar
\end{aligned}
$$

Thus, the de Broglie wavelength provides an explanation of the quantization condition on the angular momentum that Bohr needed to apply his model to explain the Balmer series. The de Broglie hypothesis forces a rethinking of the structure of matter from tiny little particles to waves. De Broglie's hypothesis that all particles have wave properties was a bold one, in that, at the time he made it, there was no direct experimental evidence supporting it. Today, technology has advanced to the point that a direct test of the hypothesis that electrons have wave properties is possible. Diffraction of a beam of electrons, similar to the double-slit experiment, has supported the wave theory of matter along with the success of quantum mechanics.

28.5   QUANTUM MECHANICS

The underlying concept of Quantum Mechanics is conservation of energy of the electron wave, which can be represented in equation (28.10) as

$$K + U = E \tag{28.10}$$

where K, U, and E represent kinetic energy, potential energy, and the total energy, respectively. Schrödinger Equation, which is a fundamental equation of Quantum Mechanics starts with the standard form of kinetic energy, equation (16.6) $K = \frac{1}{2}m_e v^2$ of the electron from Chapter 16, where m_e is the mass of the electron, $m_e = 9.11 \times 10^{-31}$ kg. Substituting into this expression for the momentum, equation (17.2), results in an expression for the kinetic energy of the electron in equation (28.11) as

$$K = \frac{p^2}{2m_e} \tag{28.11}$$

From the de Broglie wavelength expression, equation (28.9), the momentum of the electron is

$$p = \frac{h}{\lambda_D}$$

so, the kinetic energy of the electron can be written as

$$K = \frac{h^2}{2m_e \lambda_D^2}$$

$$K = \left(\frac{\hbar^2}{2m_e}\right)\left(\frac{2\pi}{\lambda_D}\right)^2$$

Therefore, equation (28.10), the energy of the electron wave becomes the expression in equation (28.12)

$$\left(\frac{\hbar^2}{2m_e}\right)\left(\frac{2\pi}{\lambda_D}\right)^2 + U = E \tag{28.12}$$

Since the electron is a wave, it can be defined as a *wave-function of the electron* (Ψ) in equation (28.13) as

$$\Psi = \Psi_o sin\left(\frac{2\pi}{\lambda_D}x\right) \tag{28.13}$$

This is similar to the representation of the waves studied previously in Chapter 15: this wave oscillates and moves from left to right. The major difference is that this wave function is measured with respect to the wavelength and not the period. So, since the sine function repeats itself every 2π, when the value of x matches, a wavelength the value of sine returns to the initial value (see Figure 28.8).

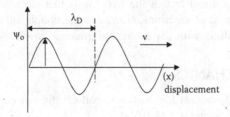

FIGURE 28.8   Wave function of an electron.

The next step in the description requires a procedure in Calculus called a second derivative. If a second derivative of the wavefunction defined in equation (28.13) is taken, the solution is

$$\frac{d^2\Psi}{dx^2} = -\left(\frac{2\pi}{\lambda_D}\right)\Psi_o sin\left(\frac{2\pi}{\lambda_D}x\right)$$

Putting this second derivative together with equation (28.12), the expression of Conservation of Energy, gives the Schrödinger's Equation in equation (28.14) as

$$-\left(\frac{\hbar^2}{2m_e}\right)\frac{d^2\Psi}{dx^2} + U\Psi = E\Psi \qquad (28.14)$$

which is the fundamental expression for the wave interpretation of Quantum Mechanics.

The process of quantum mechanics requires a first step of developing the expression for the potential energy, U, of the physical system to be analyzed. The potential energies can be as simple as zero for an electron that moves freely in a region, or the potential energy due to electrostatic attraction of positive and negative charges in an atom $U = -Fr = -ke^2/r$, or even the potential energy of a molecular vibration, which has the potential energy similar to that of a spring, $U = -1/2 k x^2$, where k is the spring constant of the electron attached to the molecule. Because calculus is not used in this text, the process of employing Schrödinger's Equation will not be explored, but some of the results will be introduced in the following section.

The simplest example is that of an electron moving freely in a region, with the potential energy, $U = 0$, for the electron at zero. If this expression for $U = 0$ is inserted into Schrödinger's equation, the expression of the energy of the electron is produced

$$-\left(\frac{\hbar^2}{2m_e}\right)\frac{d^2\Psi}{dx^2} = E\Psi$$

The Schrodinger equation is solved by assuming an answer of the form of equation (28.12). When the derivative is taken twice, the Schrödinger's equation results in the energy of the electron as

$$\left(\frac{\hbar^2}{2m_e}\right)\left(\frac{2\pi}{\lambda_D}\right)^2 = E$$

So, the energy of an electron is just the kinetic energy.

$$E = \left(\frac{\hbar^2}{2m_e}\right)\left(\frac{2\pi}{\lambda_D}\right)^2 = \left(\frac{h^2}{2m_e}\right)\left(\frac{2\pi}{\lambda_D}\right)^2 = \left(\frac{1}{2m_e}\right)\left(\frac{h^2}{\lambda_D^2}\right) = \left(\frac{1}{2m_e}\right)\left(\frac{h}{\lambda_D}\right)^2 = \left(\frac{p^2}{2m_e}\right)$$

$$E = \frac{1}{2}m_e v^2$$

So, for this simple situation, Schrödinger's equation reduces down to the kinetic energy of the electron.

28.5.1 Quantum Dot

An example for which the energy of a free electron is important is the *quantum dot*, which is a tiny group of atoms that restricts the location of the electron wave. The quantum dots that are being fabricated for use in new televisions are approximately 1–10 nm in diameter, which corresponds to 10–50 atoms wide (see Figure 28.9).

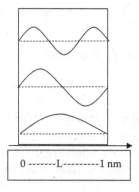

FIGURE 28.9 Wave function in the quantum dot energy levels.

In one dimension, this can be thought of as a potential that is zero in a confined area and infinitely strong outside the area, so the electron wave is confined to the area of the dot.

So, inside the quantum dot, the electron has energies that are

$$E = \left(\frac{h^2}{2m_e}\right)\left(\frac{1}{\lambda_D^2}\right)$$

but since the electron waves are trapped by the dot the wavelengths that can fit on the dot follow the rules of standing waves, as given here as

$$\frac{n}{2}\lambda_D = L, \quad \text{with } n = 1, \ 2, \ 3, \ ....$$

So, the *wavelengths of the electrons* are dictated by the size of the quantum dot, as:

$$\lambda_D = \frac{2L}{n}, \quad \text{with } n = 1, \ 2, \ 3, \ ....$$

Thus, for a quantum dot that restricts the electron wave to a space of L the energy of each level is:

$$E_n = n^2\left(\frac{h^2}{2m_e}\right)\left(\frac{1}{4L^2}\right)$$

If the quantum dot has a diameter of $L = 0.697$ nm, the energy of each levels is:

$$E_n = n^2\left(\frac{h^2}{2m_e}\right)\left(\frac{1}{4L^2}\right) = n^2\left(\frac{(6.63 \times 10^{-34} Js)^2}{2(9.11 \times 10^{-31} kg)}\right)\left(\frac{1}{4(0.697 \times 10^{-9} m)^2}\right)$$

$$E_n = n^2(1.24 \times 10^{-19} J)$$

So, the difference in the energy of the second and first level of the quantum dot is:

$$\Delta E_n = E_2 - E_1 = [(2)^2 - (1)^2]\ 1.24 \times 10^{-19} J = (3)(1.245 \times 10^{-19} J) = (3.72 \times 10^{-19} J).$$

Thus, a photon produced by an electron making a transition between the second and first levels will produce a photon with a frequency $v = (\Delta E_n)/h = 5.614 \times 10^{14}$ Hz and a photon wavelength of

$$\lambda = \frac{c}{v} = 534 \times 10^{-9}\,m = 534\ nm,$$

which is bright green. This is the size of the quantum dots that are being made for TV sets.

28.5.2 THE HYDROGEN ATOM

The next example is that of the orbiting electron of the hydrogen atom. In this case, the electron is attracted by a potential energy of the form: $U = \frac{-ke^2}{r}$

If this expression for U is inserted into Schrödinger's equation, the expression of the energy levels of the atom is produced by solving the Schrödinger equation

$$-\left(\frac{\hbar^2}{2m_e}\right)\frac{d^2\Psi}{dx^2} + \frac{-ke^2}{r}\Psi = E\Psi$$

in three dimensions. The best choice of coordinate systems is spherical coordinates (Figure 28.10), where x is replaced (r, θ, ϕ) so the equation gets very complicated very quickly. The solution is generated by separating the variables so that the wave function (Ψ) is represented by the product: $\Psi(r,\phi,\theta) = R(r)\Phi(\phi)\Theta(\theta)$.

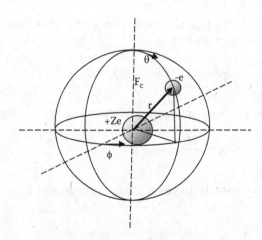

FIGURE 28.10 The 3D spherical coordinate labels of the electron in the hydrogen atom.

The separation leads to three equations for the three spatial variables of radius (r), azimuthal angle (ϕ), and zenith angle (θ). The three equations are linked together and can be solved if and only if three restrictions are met. These restrictions are the rules for the combination of electrons in the levels of atoms. The restriction of the radial equation is that the values of n (principle quantum number) are:

$$n = 1, \quad 2, \quad 3, \quad 4, \quad ....$$

The restriction of the azimuthal equation is that the values of l (orbital quantum number) are:

$$\ell = 0, 1, 2, 3, ... (n - 1).$$

The restriction of the zenith equation is that the values of m_l (magnetic quantum number) are:

$$m_\ell = -\ell, -\ell + 1, ..., 0, ... \ell.$$

These restrictions on the solution of the quantum numbers, the spin of the electron as either up ($m_s = +\frac{1}{2}$) or down ($m_s = -\frac{1}{2}$), and the Pauli exclusion principle, which explains that each electron occupies its own state defined by the quantum numbers (n, ℓ, m_ℓ, m_s), gives the electron structure of the elements of the periodic table.

The shell structures of atoms follow the following arrangement:

If $n = 1$, $\ell = 0$, $m_l = 0$, $m_s = +\frac{1}{2}$

$\ell = 0$, $m_l = 0$, $m_s = -\frac{1}{2}$

There are two possible states for electrons in the $l = 0$ shell, which is called the s shell.

If $n = 2$, $\ell = 0$, $m_l = 0$, $m_s = +\frac{1}{2}$

$\ell = 0$, $m_l = 0$, $m_s = -\frac{1}{2}$
$\ell = 1$, $m_l = -1$, $m_s = +\frac{1}{2}$
$\ell = 1$, $m_l = -1$, $m_s = -\frac{1}{2}$
$\ell = 1$, $m_l = 0$, $m_s = +\frac{1}{2}$
$\ell = 1$, $m_l = 0$, $m_s = -\frac{1}{2}$
$\ell = 1$, $m_l = +1$, $m_s = +\frac{1}{2}$
$\ell = 1$, $m_l = +1$, $m_s = -\frac{1}{2}$

There are two possible states for electrons in the $\ell = 0$ and six more in the $\ell = 1$ shell, which is called the p shell. So, there are a total of eight states in the $n = 2$ shell.

You may have heard of the labels for the values of ℓ as $(1, 2, 3, 4, ...) = $ (s, p, d, f, ...).

Example

Find the ground state electron configuration of carbon-12 ($_6C^{12}$), which is the most common isotope of carbon found in nature. Since carbon has six electrons in its balanced configuration the shells filled are:

$n = 1$, $\ell = 0$, $m_l = 0$, $m_s = +\frac{1}{2}$

$\ell = 0$, $m_l = 0$, $m_s = -\frac{1}{2}$

$n = 2$, $\ell = 0$, $m_l = 0$, $m_s = +\frac{1}{2}$

$\ell = 0$, $m_l = 0$, $m_s = -\frac{1}{2}$
$\ell = 1$, $m_l = -1$, $m_s = +\frac{1}{2}$
$\ell = 1$, $m_l = 0$, $m_s = +\frac{1}{2}$

So, the configuration is $1s^2\ 2s^2\ 2p^2$.

Which of the six 2p states are filled with an electron is not known, since they all have the same energy, but Hund's rule tells us that they at least have different m_l values.

28.6   ANSWER TO THE CHAPTER QUESTION

The "orbitals" shown in most chemistry books and sketched in Figure 28.1 are not observations, but are plots of the numerical solutions of the Schrödinger Equation with a Coulomb potential. So, when you think of an electron making transitions from the third to the second level and emitting a red photon at 657 nm, in the process one of the electrons in the atom is actually just changing in shape and not jumping from one level to another. At least that is the quantum mechanical way of looking at the situation.

28.7   QUESTIONS AND PROBLEMS

28.7.1   MULTIPLE CHOICE QUESTIONS

1. In the Bohr model of the hydrogen atom, the distance between adjacent orbits:
    A.   is the same for all orbits.
    B.   increases as the distance from the proton increases.
    C.   decreases as the distance from the proton increases.
2. In the Bohr model of the hydrogen atom, as the electron moves from an orbit further from the proton to an orbit closer to the proton, the value of its kinetic energy:
    A.   must become a negative number with a greater magnitude.
    B.   must become a negative number with a smaller magnitude.
    C.   must become a positive number with a greater magnitude.
    D.   must become a positive number with a smaller magnitude.
    E.   remains constant.
3. As compared to the photon generated when an electron transitions from the forth to the third level of a hydrogen atom, the photon generated when an electron transitions from the second to the first level has
    A.   more energy.
    B.   less energy.
    C.   the same energy.
4. Is the wavelength of the photon produced in the transition from the third to the first orbit of the electron in hydrogen, longer than visible (infrared), in the visible, or shorter than visible (ultraviolet) part of the e&m spectrum?
    A.   longer (infrared)
    B.   visible
    C.   (shorter) ultraviolet
5. If photon A has a wavelength that is longer than the wavelength of photon B, which photon, A or B, has the greater energy?
    A.   photon A
    B.   photon B
    C.   neither – both energies are equal

6. Is the energy of the photon produced from the transition of an electron from the fourth to the second orbit in hydrogen greater than, less than, or equal to the energy of the photon produced from the transition of an electron from the third to the first orbit in hydrogen?
   A. greater than
   B. less than
   C. equal to

7. How many electron wavelengths (de Broglie wavelength) are there of the electrons orbiting in the third orbit of hydrogen?
   A. 1
   B. 3
   C. 9
   D. 18

8. Is the wavelength of the photon produced in the sixth to third transition of the electron in singly ionized Helium (+He) longer than, shorter than, or equal to the wavelength of the photon produced from the sixth to third transition of the electron in hydrogen?
   A. longer than
   B. shorter than
   C. equal to

9. For the shell with the principle quantum number of 2, how many quantum states are available for electrons to occupy?
   A. 2
   B. 4
   C. 6
   D. 8
   E. 10
   F. 12

10. What is the highest principle quantum number, n, of the quantum state (orbit) with at least one electron in it for a ground state sodium ($_{11}Na^{22}$) atom?
   A. 1
   B. 2
   C. 3
   D. 4
   E. 5
   F. 11

28.7.2   PROBLEMS

1. Check that the expressions for ground state energy and the Bohr radius yield the correct units.
2. Compute the energy of an electron in the third and fifth states of hydrogen.
3. Compute the wavelength of the photon produced when an electron makes a transition from the fifth to the third state of hydrogen.
4. Find r_n for the case $n = 20,000$.
5. Calculate the classical frequency of the electron in the $n = 20,000$ orbit.
6. Calculate the frequency of the light given off in the transition from $n = 20,000$ to $n = 19,999$ using the Bohr model.
7. Compute the wavelength of the photon produced when an electron makes a transition from the third to the first electron obit of singly ionized Helium (+He).
8. Compute the wavelength of the photon produced when an electron makes a transition from the sixth to the third electron obit of hydrogen.

9. a. Compute the "electron wavelengths" in the third shells of hydrogen.
   b. Sketch the "electron waves" of the third shell around orbits in hydrogen.
10. What is the ground state electron configuration of Neon ($_{10}Ne^{20}$)?

Table of elementary particle masses and charges from Chapter 1.

Particle	Charge (C)	Mass (kg)
Electron	-1.602×10^{-19}	9.11×10^{-31}
Proton	$+1.602 \times 10^{-19}$	1.672×10^{-27}
Neutron	0	1.674×10^{-27}

29 Quantum Nature of Light

29.1 INTRODUCTION

The quantum nature of light is the topic studied in this chapter. Because of its scientific and historical significance, the chapter begins with an explanation of photoelectric effect. Albert Einstein used a quantum-based explanation of the photoelectric effect to help establish the concept of the photon, which is a particle of light. In addition, the understanding of the photoelectric effect provides an underlying structure to understand the generation and the characteristics of X-rays. Next, Compton scattering of X-rays of electrons is presented as a momentum-based example of the quantum nature of light. The chapter concludes with a brief introduction to the laser and includes a more extensive overview of the laser in an appendix. This appendix provides an opportunity to synthesize many of the topics covered in this volume, while explaining the key parts of this important quantum-electronic device.

29.2 CHAPTER QUESTION

The word LASER is an acronym that stands for Light Amplification by Stimulate Emission of Radiation. So what is stimulate emission, and why is it central to the laser? To answer this question, an understanding of the quantum nature of light is required. After this topic is covered throughout this chapter, this question will be answered at the end of the chapter. In addition, there is an appendix to this chapter, which offers a more in-depth introduction to the concept of lasers and provides an opportunity to apply many of the topics covered in this volume.

29.3 THE CATHODE RAY

The place to start an understanding of the quantum nature of light is the device used to first discover the quantum nature of charge. In the late 1800s, J.J. Thompson used a cathode ray tube (CRT) to discover the electron, which is a particle that is the quantized negative electrical charge. The CRT, depicted in Figure 29.1, is a sealed glass tube with electrodes on each end.

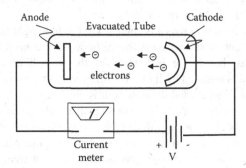

FIGURE 29.1 A cathode ray tube (CRT).

The tube is evacuated, and a high voltage is put across electrodes in the tube.

If the voltage across the tube is high enough, the electrons will be pushed off the cathode, and they will accelerate across the empty tube to the anode. The flow of charge will continue around the circuit, registering a current through the meter. Using a system similar to a velocity selector in a

DOI: 10.1201/9781003308072-30

mass spectrometer, Thompson crossed electric and magnetic fields to find the charge-to-mass ratio of the electron.

29.4   PHOTOELECTRIC EFFECT

Another early experiment used to demonstrate the quantum nature of light employed a cathode ray tube and a monochromatic, adjustable light source. If the CRT voltage is set too low, the electric field between the cathode and anode will not be strong enough to liberate electrons from the cathode to travel across the tube to the anode, and no current will flow through the circuit. Without adjusting the voltage, if light of different frequencies is shown onto the cathode, as shown in Figure 29.2, a current will flow at some frequencies of light but not others.

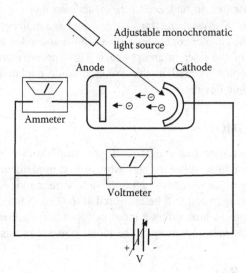

FIGURE 29.2   Photoelectric effect device.

One of the most puzzling results was that if a frequency of light shown onto the cathode did not cause a current to flow, increasing the intensity of the light did not lead to a current. On the other hand, if a certain frequency or color of light shown onto the cathode resulted in a flow of current, increasing the intensity of the light increased the current.

For example, a current will flow across a particular CRT if blue light is shown onto the cathode, but will not flow if red light is shown onto the cathode. If the brightness of the red light is increased, nothing happens, but if the brightness of the blue light is increased, the current increases. Since this current is initiated by light, it is referred to as a *photocurrent*. It is still just the flow of electrical charge, but light started the flow so it is given this name. This experimental result puzzled the scientific community since they were sure that light acted only as a wave, and the energy of a wave is proportional to the amplitude of the wave. Since, we perceive the amplitude of the wave as the brightness of the light, it was assumed that if the intensity is increased, the energy of the wave is increased, so everyone thought eventually a current should flow, but it did not.

In 1905, Albert Einstein applied Planck's black body theory, which states that the energy (E) of the oscillators in a heated solid are quantized with a value of

$$E = hf$$

where h is Planck's constant: $h = 6.626 \times 10^{-34}$ J s and f is the frequency of the oscillators. Einstein took this theory further to explain, in his 1921 Nobel Prize winning paper titled a "Concerning a

Heuristic Point of View Toward the Emission and Transformation of Light", that light is also quantized. These quantized particles of light are now called photons, and they have an energy (E_p) equal to the product of the frequency, v, of the light and Planck's constant, h, which was introduced as equation (25.4) and rewritten here for completion as,

$$E = hv$$

Note that there is a switch of the variable for frequency from f for the frequency of the oscillators that produced the light and v for the frequency of the generated light. These frequencies will have the same value, but are two different frequencies, so the switch is made to make this distinction clear.

Using the concept of the quantized particle of light, Einstein explained the photoelectric effect as follows. The maximum kinetic energy (K_{max}) an electron can have in the photoelectric effect experiment is equal to the energy of the photon minus any energy that is used as work to free the electron from the metal. This concept is expressed in equation (29.1) as

$$K_{max} = E_p - \phi, \tag{29.1}$$

where ϕ is the work function of the metal and is equal to the energy needed to remove electrons from the surface of the metal. It is common to rearrange this equation and substitute hv for E_p to get equation (29.2) as

$$hv = K_{max} + \phi. \tag{29.2}$$

During the photoelectric effect experiment, photons with energy hv are incident upon the cathode of a vacuum tube, and these electrons in the cathode use energy from the photons to escape the surface with a maximum energy of K_{max}. The emitted electrons reach the anode of the tube and are measured as the photoelectric current. The critical concept is that each electron can only absorb one photon at a time. Thus, the frequency of the photon, or its color, determines if it has enough energy to liberate an electron from the metal cathode.

The intensity of the light is related to the number of photons, and the color of the light is related to the energy of the photons in the light. Thus, if a beam of red light is shown onto the cathode of the photoelectric effect and no photocurrent is registered, then increasing the brightness will not help, since it simply delivers more photons each with too little energy to liberate an electron. On the other hand, if violet light has a high-enough frequency to liberate electrons, then turning up the intensity will cause the photocurrent to increase since more electrons are liberated per second.

Example 1

Calculate the longest wavelength of light that will cause a photocurrent in a photoelectric effect experiment if the cathode is made of iron, which has a work function of $\phi_{Fe} = 4.50$ eV.

Solution

Given that the work function of iron is: $\phi_{Fe} = 4.50$ eV, the work function can be converted to Joules by simply multiplying by the conversion factor.

$$\phi_{Fe} = (4.50 \ eV) \ (1.602 \times 10^{-19} \ J/eV) = 7.209 \times 10^{-19} \ J$$

Set $K_{max} = 0$ to find the photon energy (E_p), which will just overcome the work function to liberate the electrons of the cathode. Thus, equation (29.2) becomes:

$$hv = \phi_{Fe} = 7.209 \times 10^{-19} \, J$$

Solving for the frequency of the light gives,

$$v = \frac{\phi_{Fe}}{h} = \frac{(7.209 \times 10^{-19} \, J)}{(6.626 \times 10^{-34} \, Js)} = 1.088 \times 10^{15} \, Hz$$

To find the wavelength, use equation (25.3) solved for the wavelength,

$$\lambda = \frac{c}{v} = \frac{3 \times 10^8 \, \frac{m}{s}}{1.088 \times 10^{15} \, Hz} = 2.76 \times 10^{-7} \, m = 276 \ nm$$

This is an ultraviolet photon.

Example 2

Calculate the maximum kinetic energy of an electron ejected by an ultraviolet photon with a wavelength of 200 nm, shown upon an aluminum ($\phi_{Al} = 4.08$ eV) cathode in a photoelectric tube.

Solution

Find the frequency of the light using equation (25.3), since the wavelength gives,

$$v = \frac{c}{\lambda} = \frac{3 \times 10^8 \, \frac{m}{s}}{200 \times 10^{-9} \, m} = 1.5 \times 10^{15} \, Hz$$

Find the photon energy with equation (25.4) as,

$$E_p = hv = (6.626 \times 10^{-34} \, J \, s)(1.5 \times 10^{15} \, Hz) = 9.939 \times 10^{-19} \, J$$

Convert the work function from eV to J as follows,

$$\phi_{Al} = (4.08 \ eV)(1.602 \times 10^{-19} \, J/eV) = 6.536 \times 10^{-19} \, J$$

Rearrange equation (29.2) and plug in the values to find the maximum kinetic energy of the photoelectrons,

$$K_{max} = E_p - \phi_{Al} = 9.939 \times 10^{-19} \, J - 6.536 \times 10^{-19} \, J = 3.403 \times 10^{-19} \, J = 2.12 \ eV$$

29.4.1 Stopping Voltage

Another common arrangement of the photoelectric effect experiment, shown in Figure 29.3, is used to find the stopping voltage (V_s) for a particular frequency of light and cathode material.

The arrangement looks very similar to the one shown in Figure 29.2, except the (+) and (−) voltages are switched on the battery. The experiment starts with zero voltage across the CRT, so that the adjustable voltage supply is set to zero, and a light source with a high-enough frequency, or short-enough wavelength, to liberate electrons is shown onto the cathode metal. Some of these liberated electrons make their way across the tube to the anode. The cathode becomes positive,

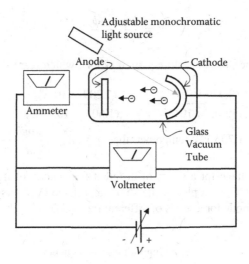

FIGURE 29.3   Photoelectric apparatus for stopping voltage.

since it lost electrons, and a current begins to flow through the circuit. The voltage (V) is increased until the current stops. This is the stopping voltage (V_s), since it stops electrons from crossing the tube in the first place by drawing them back to the cathode.

Since the voltage is energy per charge, the product of the stopping voltage and the charge of the electrons it is stopping is an energy that must be equal in magnitude to the maximum kinetic energy, K_{max}, of the electrons to stop them from becoming part of the current that is stopped. Thus, equation (29.2) for the stopping voltage experiment becomes equation (29.3) as

$$h\nu = V_s e + \phi, \tag{29.3}$$

Example 3

Calculate the stopping voltage for an ultraviolet photon with a wavelength of 200 nm shown upon an iron (ϕ_{Fe} = 4.50 eV) cathode of a photoelectric experiment set up to measure the stopping voltage, as shown in Figure 29.3.

Solution

Convert the work function of iron into Joules:

$$\phi_{Fe} = (4.50 \ eV) \ (1.602 \times 10^{-19} \ J/eV) = 7.209 \times 10^{-19} \ J$$

Find the frequency of the light using equation (25.3) since the wavelength is given,

$$\nu = \frac{c}{\lambda} = \frac{3 \times 10^8 \ \frac{m}{s}}{200 \times 10^{-9} \ m} = 1.5 \times 10^{15} \ Hz$$

Find the photon energy with equation (25.4) as,

$$E_p = h\nu = (6.626 \times 10^{-34} \ J \ s)(1.5 \times 10^{15} \ Hz) = 9.939 \times 10^{-19} \ J$$

Rearrange equation (29.3) to find the stopping voltage as equation (29.4) as,

$$V_s = \frac{h\nu}{e} - \frac{\phi}{e} \qquad (29.4)$$

Plug in the values from the problem into equation (29.4), which results in:

$$V_s = \frac{(9.939 \times 10^{-19} J)}{(1.602 \times 10^{-19} C)} - \frac{(7.209 \times 10^{-19} J)}{(1.602 \times 10^{-19} C)}$$

$$V_s = 1.704 \; V, \; \text{remember that a Volt} = \frac{Joule}{Coulomb} = \frac{J}{C}$$

Measuring the stopping voltage for a series of different frequencies of light provides a method to compute Planck's constant. First, a plot of the stopping voltage (V_s) versus the frequencies of the light, ν, illuminated the cathode for a series of different frequencies of light, is plotted in Figure 29.4.

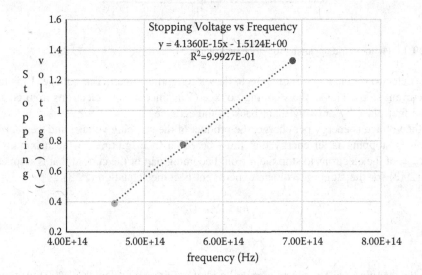

FIGURE 29.4 Plot of stopping voltage (V_s) vs the frequency (f) of the light shown upon the cathode and a line fit to the data points.

Next, a line is fit to the data, and the slope and y-intercept of the best fit line are recorded as:

$$m = slope = 4.136 \times 10^{-15} \frac{V}{Hz}$$

and

$$b = \text{y-intercept} = 1.512 \; \overset{.}{V}.$$

Comparing the standard equation of a linear fit, $y = mx + b$, to the stopping voltage equation (29.4) $V_s = \left(\frac{h}{e}\right)\nu - \left(\frac{\phi}{e}\right)$ reveals that the slope is equal to $\left(\frac{h}{e}\right)$ and the y-intercept is equal to $-\left(\frac{\phi}{e}\right)$. First, find h from the slope:

$$\frac{h}{e} = slope = 4.136 \times 10^{-15} \frac{V}{Hz}$$

$$h = (slope)(e) = \left(4.136 \times 10^{-15} \frac{V}{Hz}\right)(1.602 \times 10^{-19} C) = 6.626 \times 10^{-34} \; Js$$

Unit check: $\left(\frac{V}{Hz}\right)(C) = \left(\frac{J/C}{1/s}\right)(C) = Js$

Next, use the y-intercept, which is equal to $\frac{\phi}{e}$, to find the work function for this experiment:

$$\left(\frac{\phi}{e}\right) = \text{y-intercept} = 1.512 \text{ V}$$

$$\phi = 1512 \text{ V} * 1.602 \times 10^{-19} \text{ C} = 2.422 \times 10^{-19} \text{ J} = 1.512 \text{ eV}$$

29.5   X-RAYS

X-rays are produced by bombarding a metal target with high-speed electrons. The kinetic energy of the high-speed electrons is converted to light energy in the form of X-ray photons. These photons have such a high energy they will pass through soft tissue. This is why X-rays are a key for medical imaging.

X-rays are produced in an X-ray tube, which is an evacuated glass tube containing a cathode and an anode, as depicted in Figure 29.5.

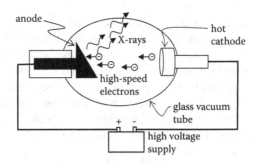

FIGURE 29.5   X-ray tube.

The cathode in an X-ray tube is at such a high negative voltage that the electrons in the metal of which the cathode is made have a large-enough kinetic energy that some of them escape the cathode in a process called thermionic emission. Remember that the work functions of metals are only a few electron Volts, so a common voltage of 20 kV (20,000 V) to 30 kV (30,000 V) across the X-ray tube is more than enough to free some electrons from the cathode. The electrons that do escape the cathode are accelerated across the tube, away from the negative cathode and toward the positive anode. The electrons gain a high kinetic energy as they race across the tube. X-rays are produced when the electrons crash into the metal atoms of the anode. Remember that e&m radiation is generated by accelerated charged particles, and because the electrons have such a high kinetic energy just before they collide with the metal atoms, the e&m radiation resulting from the massive deceleration due to the collisions of the electrons with the anode have a maximum frequency in the X-ray part of the electromagnetic spectrum. That is, the energy of the photons resulting from a collision can be any amount from 0 up to the kinetic energy of the incoming electron, depending on the exact nature of the collision of the electron with the anode. The X-rays produced by the deceleration of the electrons is called *bremsstrahlung*, which is German word meaning "braking radiation." Considering the X-ray spectrum depicted in Figure 29.6 to consist of a main curve plus two sharp peaks, the main curve represents the intensity of the bremsstrahlung as a function of wavelength.

The minimum wavelength of the bremsstrahlung curve corresponds to the X-ray with highest possible energy. By combining equation (25.3), $c = \lambda v$, and equation (25.4), $E = hv$, into equation (29.5) as

$$E_\gamma = \frac{hc}{\lambda} \tag{29.5}$$

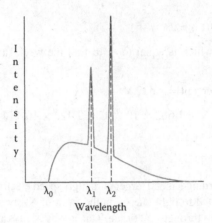

FIGURE 29.6   X-ray spectrum.

it is clear that the energy of a photon (E_γ) is inversely proportional to the wavelength (λ) of the photon. Therefore, the X-ray with the highest energy has the same energy as the kinetic energy of the incoming electron. Compared to the energy gained by the electrons in being accelerated by the electric field between the cathode and the anode, the kinetic energy of the electrons escaping the cathode by thermionic emission is negligible, so the kinetic energy of the electrons when they reach the anode, is equal to the absolute value of the change in their potential energy as they go from cathode to anode. That is, since voltage is potential energy (U) per charge, then the potential energy is just the product of voltage (V) and charge of the electron (q_e), as shown in the first part of equation (29.6),

$$|\Delta U| = |q_e|V \tag{29.6}$$

and this potential energy is converted to kinetic energy of the electron (K_e) as it accelerates across the X-ray tube, as shown in the expression of conservation of energy in equation (29.7)

$$|\Delta U| = K_e \tag{29.7}$$

Example 1

Compute the shortest wavelength (λ_{min}) of X-rays produced in an X-ray tube operating at a voltage of 20 kV.

First, find the kinetic energy of the electrons arriving at the anode by combining equations (29.6) and (29.7) to get,

$$K_e = |q_e|V = (1.6022 \times 10^{-19} \text{ C}) (20,000 \text{ V})$$
$$K_e = 3.2044 \times 10^{-15} \text{ J}$$

Next, the energy of the highest energy photon is set equal to the kinetic energy of the electron,

$$E_\gamma = K_e = 3.2044 \times 10^{-15} \text{ J}$$

The final step is finding the minimum wavelength X-ray, which is the highest energy X-ray, using equation (29.5). This can be done in two steps or just one. First, in two steps, find the frequency of the photon using, $E_\gamma = h\nu$, as

$$\nu = \frac{E_\gamma}{h} = \frac{3.2044 \times 10^{-15} \, J}{6.626 \times 10^{-34} \, Js} = 4.836 \times 10^{18} \, Hz.$$

and then find the wavelength using, $c = \lambda \nu$, as

$$\lambda_o = \frac{c}{\nu} = \frac{3 \times 10^8 \, \frac{m}{s}}{4.836 \times 10^{18} \, Hz} = 6.1992 \times 10^{-11} \, m = 0.061992 \; nm.$$

The alternative is to find the wavelength in one step, by rearranging equation (29.5) as

$$\lambda_o = \frac{hc}{E_\gamma} = \frac{(6.626 \times 10^{-34} \, Js)\left(3 \times 10^8 \, \frac{m}{s}\right)}{3.2044 \times 10^{-15} \, J} = 6.1992 \times 10^{-11} \, m = 0.061992 \; nm$$

In both cases, the wavelength is labeled λ_o to match the label of the minimum wavelength in Figure 29.6.

The X-rays at the wavelengths of the two sharp peaks in the system are referred to as *characteristic X-rays,* because they are characteristic of the metal of which the anode is made. In the early 1900s, Henry Moseley conducted one of the first systematic studies of the X-ray spectrum. Using an X-ray tube to produce X-rays, Mosley adjusted the voltage of the anode relative to the cathode. When he increased the voltage, the minimum wavelength of the curve decreased and the curve expanded, but the wavelengths of the characteristic X-rays, the spikes in the spectrum, did not change. When Mosley changed the metal of which the anode was made, the wavelengths of characteristic X-rays changed. A summary of the data similar to what he produced is given in Table 29.1.

TABLE 29.1

Data of Characteristic X-Rays for Different Anode Metals

Element	Atomic Number	Atomic Mass [u]	Wavelength of Spike 1 (Kβ) [nm]	Wavelength of Spike 2 (Kα) [nm]
Manganese (Mn)	25	54.94	.191	.210
Iron (Fe)	26	55.84	.175	.193
Cobalt (Co)	27	58.93	.162	.179
Nickel (Ni)	28	58.69	.150	.166
Copper (Cu)	29	63.55	.139	.154

It is clear from the data that the wavelengths of the characteristic X-rays decrease as the atomic number of the anode material increases. Mosley correctly deduced that the high-speed electrons collide with electrons in the lowest-energy orbit of the anode atoms, dislodging electrons in that orbit, which is closest to the nucleus of the atom. The atomic number indicates the number of protons in the nucleus, and more protons attract the electrons more strongly. Thus, when electrons from higher-energy orbits transition down, as depicted in Figure 29.7, to the open location in the lowest-energy orbit of these atoms, there is a release of energy in the form of an X-ray photon.

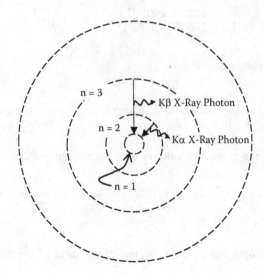

FIGURE 29.7   Transitions that produce characteristic X-ray photons.

The more protons there are in the nucleus of the atom, the more negative the energy of the lowest-energy state, the $n = 1$ state, and the greater the energy difference between it and the $n = 2$ state, as well as between it and the $n = 3$ state. Thus, the greater the atomic number, the greater the energy of the photons released, meaning the shorter the wavelengths of the photons. This logic is the cornerstone of the understanding of the characteristic "spikes" of the X-ray spectrum.

29.6   COMPTON SCATTERING

Another important example of the particle nature of light is the scattering of X-rays by electrons in a solid. When X-rays are directed onto a solid material, such as a piece of carbon, they scatter off in many directions, which is the case when most electromagnetic waves are directed at a solid.

In the case of the scattered X-rays off a solid, the wavelength of some of the scattered X-rays is changed in the scattering process. This type of scattering is known as *Compton scattering* because it was explained by Arthur H. Compton by using conservation of momentum in 2-D, using conservation of energy, and treating the X-rays as particles. Studying the case of X-rays scattered off solid carbon, Compton reasoned that some of the X-rays were scattering off electrons in outer orbitals of the carbon atoms, where the electrons are so weakly bound to the carbon atom that, in the collisions, the electrons behave as free electrons. Compton reasoned that the scattering event reduces the energy of the X-ray and thus increases the wavelength of the X-ray.

He took this idea to its logical next step, and applied conservation of momentum and conservation of energy to the collision, depicted in Figure 29.8. He solved for the wavelength of the light after the scattering event, thus deriving what is now known as the Compton scattering formula, given in equation (29.8) as

$$\lambda_f = \lambda_i + \frac{h}{m_e c}(1 - \cos\theta) \tag{29.8}$$

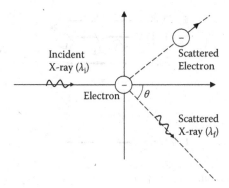

FIGURE 29.8   Compton scattering.

where:
  θ is the angle between the original direction of travel of the X-ray photon and the direction of travel of the X-ray photon after the collision.
  λ_i is the wavelength of the incoming photon
  λ_f is the wavelength of the outgoing photon

Note that, in applying conservation of momentum as part of the derivation of the Compton scattering formula, Compton used the de Broglie expression $\lambda = h/p$ to arrive at the momentum $p = h/\lambda$ for the X-ray photon. He treated the photon as a particle having energy $E_\gamma = hf$ and momentum $p = h/\lambda$.

Example 2

An X-ray of wavelength .20,000 nm is incident on an outer-orbit electron in a sample of carbon. Compute the wavelength of an X-ray scattered at an angle of 60°.

$$\lambda_f = \lambda_i + \frac{h}{m_e c}(1 - \cos \theta)$$

$$\lambda_f = .2 \text{ nm} + \frac{6.6261 \times 10^{-34} \text{ J·s}}{(9.1094 \times 10^{-31} \text{ kg})(3 \times 10^8 \text{ m/s})}(1 - \cos 60°)$$

$$\lambda_f = .2 \text{ nm} + 1.212 \times 10^{-12} \text{ m}$$

$$\lambda_f = .2 \text{ nm} + .00121 \text{ nm}$$

$$\lambda_f = .20121 \text{ nm}$$

The scattered X-ray has a longer wavelength than the incident X-ray.

29.7   ANSWER TO CHAPTER QUESTION

In 1917, Albert Einstein wrote a paper in which he explained that if electrons in an atom could spontaneously absorb and emit radiation, then these sources should also emit radiation in the presence of an e&m wave. He even wrote an equation that described the probability of this happening. The process he wrote about, in which the generation of a photon is triggered by another photon, is stimulated emission. This process is so fundamental to the generation of laser light that it is in the name. In this process of stimulated emission, a photon of light stimulates the emission of a photon from an electron in an energized atomic transition, as shown in Figure 29.9.

The wave with the arrow labeled with a 1 in Figure 29.9, moving from the left to right, will stimulate the emission of a photon if the energy of the incoming photon 1 is just right.

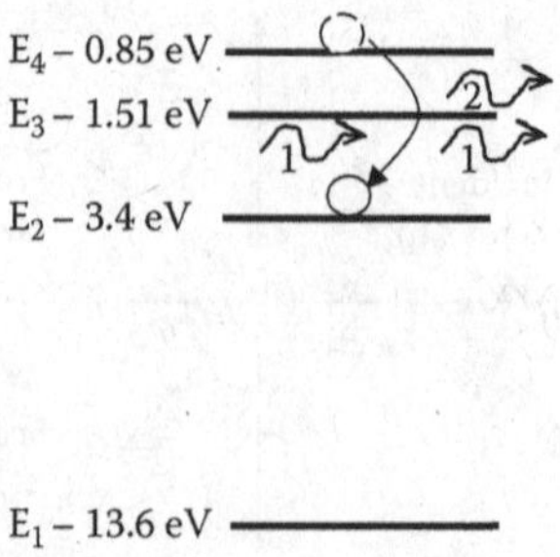

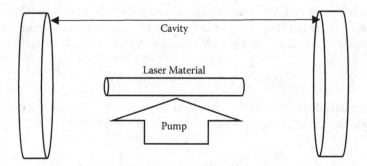

$E_4 - 0.85$ eV
$E_3 - 1.51$ eV
$E_2 - 3.4$ eV

$E_1 - 13.6$ eV

FIGURE 29.9   Stimulated emission.

In Figure 29.2, the electron in level 4 is stimulated to transition to level 3 and generates a photon, labeled with a 2, that is, its exact duplicate. This is the process of stimulated emission.

This process of stimulated emission is critical with the production of laser light, but it is important to realize that a laser is more than just the stimulated emission. The basic structure of a laser includes a cavity, a laser material, and a pump, as depicted in Figure 29.10, all of which play an important role in the device.

Cavity

Laser Material

Pump

FIGURE 29.10   The three parts of a laser, the cavity, the pump, and the laser material.

The simplest cavity is two parallel mirrors, one of which is less than 100% reflective, so some of the light can escape. The cavity keeps most of the photons generated by stimulated emission in the laser, so more amplification can result as the photons make many passes through the laser. The laser material has some specific qualities that allow it to store energy and provides more opportunities for stimulated emission. The pump is an energy source that is often a high voltage supply, or a powerful lamp, or another laser, or even a chemical reaction. This pump is critical to start the entire process. If the answer to this question "stimulates" your interest in learning more about lasers, please see the appendix of this chapter.

29.8   QUESTIONS AND PROBLEMS

29.8.1   MULTIPLE CHOICE QUESTIONS

1. If the wavelength of a photon is increased, will the energy of the photon increase, decrease, or remain the same?
   A.   increase
   B.   decrease
   C.   remain the same

2. In a photoelectric effect experiment, if light is shining on the cathode and no photo-current is produced, then can we conclude that to produce a photocurrent the wavelength of the light must be increased or decreased?
   A. increased
   B. decreased

3. In a photoelectric experiment there is no photocurrent. Which of the following changes will make it so that a photoelectric current begins to flow?
   A. Increase the intensity of the light source.
   B. Increase the wavelength of the light source.
   C. Increase the frequency of the light source.
   D. Move the light source closer to the vacuum tube.

4. In a photoelectric experiment, suppose that the variable voltage is set to 1 V, a light is shown upon the cathode, and a photoelectric current is detected. What would happen to the maximum kinetic energy with which the electrons are hitting the anode if the voltage was dialed down to .9 volts?
   A. The kinetic energy would increase.
   B. The kinetic energy would decrease.
   C. The kinetic energy of the electrons would remain the same.

5. A photoelectric experiment is set up to measure the stopping voltage, as depicted in Figure (29.3). If the frequency of the light shown upon the cathode is increased, will the stopping voltage required increase, decrease, or remain the same?
   A. increase
   B. decrease
   C. remain the same

6. A photoelectric experiment is set up to measure the stopping voltage, as depicted in Figure (29.3). The iron (ϕ_{Fe} = 4.50 eV) cathode is illuminated with light of a specific frequency, and a stopping voltage is measured as $V_{S(Fe)}$. The iron cathode is replaced with an aluminum (ϕ_{Al} = 4.08 eV) cathode that is illuminated with light of a specific frequency, and a stopping voltage is measured as $V_{S(Al)}$. Is $V_{S(Fe)} > V_{S(Al)}$, $V_{S(Fe)} < V_{S(Al)}$, or $V_{S(Fe)} = V_{S(Al)}$?
   A. $V_{S(Fe)} > V_{S(Al)}$
   B. $V_{S(Fe)} < V_{S(Al)}$
   C. $V_{S(Fe)} = V_{S(Al)}$

7. The wavelength of the characteristic X-rays of the spectrum are changed when the:
   A. anode material of an X-ray tube is changed.
   B. voltage across the X-ray tube is changed.
   C. both A and B are done.

8. The minimum wavelength of an X-ray spectrum is changed when the:
   A. anode material of an X-ray tube is changed.
   B. voltage across the X-ray tube is changed.
   C. both A and B are done.

9. In a Compton scattering experiment, an X-ray with a wavelength of λ_i is incident on a sample of carbon. If the wavelengths of the scattered X-rays measured at a scattering angle of 70° are found to be λ_f, is $\lambda_i > \lambda_f$, $\lambda_i < \lambda_f$, or $\lambda_i = \lambda_f$?
   A. $\lambda_i > \lambda_f$
   B. $\lambda_i < \lambda_f$
   C. $\lambda_i = \lambda_f$

10. In a Compton scattering experiment, an X-ray with a wavelength of λ_i is incident on a sample of carbon. The wavelength of the scattered X-rays measured at a scattering angle of 70° are found to be $\lambda_{70}°$. If the detector is moved to 50° and the measurement is

repeated so that a new wavelength of the scattered ray is found to be $\lambda_{50}°$, is $\lambda_{70}° > \lambda_{50}°$, $\lambda_{70}° < \lambda_{50}°$, or $\lambda_{70}° = \lambda_{50}°$?

A. $\lambda_{70}° > \lambda_{50}°$

B. $\lambda_{70}° < \lambda_{50}°$

C. $\lambda_{70}° = \lambda_{50}°$

29.8.2 PROBLEMS

1. Calculate the energy of a photon of blue light, with a wavelength of 450 nm.
2. Calculate the longest wavelength of light that will cause a photocurrent in a photoelectric effect experiment if the cathode is made of sodium (ϕ_{Na} = 2.28 eV).
3. Calculate the maximum kinetic energy of an electron in a photoelectric effect experiment that is ejected by an ultraviolet photon with a wavelength of 150 nm shown upon a Zinc (ϕ_{Zn} = 4.31 eV) cathode in a photoelectric tube.
4. Calculate the maximum kinetic energy of an electron ejected from a sodium surface, which has a work function of 2.28 eV when illuminated by light with a wavelength of 410 nm.
5. Calculate the stopping voltage for an ultraviolet photon with a wavelength of 150 nm shown upon a sodium (ϕ_{Na} = 2.28 eV) cathode of a photoelectric experiment.
6. Calculate the stopping voltage for a photon with a wavelength of 100 nm shown upon a Zinc (ϕ_{Zn} = 4.31 eV) cathode of a photoelectric experiment.
7. X-rays are generated in a hot-filament cathode ray tube with a high voltage of 50,000 V across the tube. What is the minimum wavelength of X-rays produced by this X-ray tube?
8. X-rays are generated in a hot-filament cathode ray tube with a high voltage of 30,000 V across the tube.
   a. Calculate the maximum kinetic energy of the electrons in the tube.
   b. Calculate the minimum wavelength of X-rays produced by this X-ray tube.
9. X-rays with a wavelength of 0.140 nm are scattered from a very thin slice of carbon. What will be the wavelength of the X-rays scattered at an angle of 40°?
10. An X-ray is incident, with a wavelength of 0.15 nm, onto a sample of carbon. Compute the wavelength of an X-ray scattered at an angle of 70°.

APPENDIX: LASER (SYNTHESIS OPPORTUNITY 2)

LASER is an acronym for **L**ight **A**mplification by **S**timulated **E**mission of **R**adiation. Although, there are many different types of lasers, they all have three main parts: a pump, a laser material, and a cavity. The pump is the source of energy that starts the laser process, known as lasing. To understand the role of each part of the laser in producing light with specific characteristics, requires a review of much of the topics covered in this volume of this textbook.

THE PUMP

The energy of the pump can come from many different sources. Some of these sources include: a continuous low-voltage electrical current, as is the case in the diode lasers found in a lot of technology; a continuous high-voltage electrical voltage, as in helium neon lasers used in barcode scanners; a pulsating bright lamp, as in some medical laser systems; a chemical process used in high-energy lasers designed to shoot down missiles; or even another laser, as is the case with many educational and research solid state and dye laser systems.

Most common lasers get their energy from the electrical grid, so they can simply be plugged into an outlet. The Helium (He) Neon (Ne) laser, known as a HeNe and pictured in Figure 29.11, is plugged into a wall outlet.

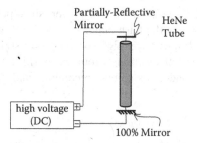

FIGURE 29.11   HeNe laser.

The power supply takes the input of the AC current from the plug and converts it to a high-voltage DC current. This high-voltage power supply, like the one used in a neon sign power supply, creates a flow of energized electrons through a tube filled with a specific ratio of helium and neon. The energized electrons collide with electrons in the heliumm and the energized helium ions give the electrons in the neon the right amount of energy to move them into a laser level. So, the helium helps the neon lase.

In the case of many other lasers, like many of those used in medical applications, the high voltage is stored in a capacitor bank and turned off and on rapidly by an electronic switch that creates a high-voltage pulse across the tube filled with gas, as depicted in Figure 29.12.

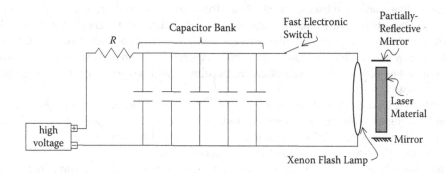

FIGURE 29.12   Flash lamp laser.

In some cases, the tube is filled with a mixture of inert gases, such as xenon, to produce an intense pulse of light. These sources of light are called "flash-lamps". The light from this lamp is used to power the laser instead of directly connecting the positive and negative electrodes across the laser material.

The high-voltage power supply charges the capacitors in the capacitor bank at a rate that is determined by the RC time constant of the circuit. Resistor, R, helps draw a current through the circuit that stores energy in the capacitor bank in the form of displaced charge. The capacitance of the circuit, C, is the sum of all the capacitances in the capacitor bank, because the capacitors are arranged in parallel so that the same voltage (V) can store energy across each of them. This works because the equivalent capacitance ($C_{eq} = C_1 + C_2 + C_3$) of parallel capacitors is the sum of the individual capacitances. Remember, this summation is the result of each capacitor having the same voltage across each capacitor, and the capacitance (C) of each capacitor is ratio of the charge (q) stored across the capacitor to the voltage (V) across the capacitor, $C = q/V$.

When the switch is triggered, all the energy stored in the capacitors is dumped across the tube. Remember, a dissipation of a large amount of energy in a short amount of time is the definition of large amount of power, $P = \Delta E/t$. Also, remember that a high voltage across and a large current through the flash lamp results in a large amount of power, since power is the product of voltage across and current through ($P = VI$) the flash lamp.

Diode lasers are pumped by low-voltage, high-current power supplies that push electrons from the negative side and holes (lack of electrons) from the positive side into the depletion region between the two types of semiconductors so they recombine and give off energy in the form of photons (Figure 29.13).

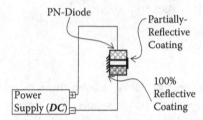

FIGURE 29.13 Diode laser.

Other mechanisms, such as chemical reactions or another laser, can also be used as the pump.

LASER MATERIAL

The qualities that make a substance a candidate for a laser material are specific, but not rare. These laser materials are solids, mostly crystals like ruby, alexandrite, and ti: sapphire, liquids, which are mostly dyes in a solvent like acetone, and gases, such as nitrogen, argon, CO_2 and other elementary and molecular gases. The only requirement is that these materials have the necessary qualities that allow for **amplification** by **stimulated** **emission** (the "**ase**" in the middle of the word **laser**).

Remember the discussion of the Bohr model of the hydrogen atom in which energized electrons transition from higher levels to lower ones to produce a bundle of electromagnetic energy called a photon, as depicted in Figure 28.6. The difference in the energy levels results in the energy of the photons and thus the wavelength of the light produced.

In producing a laser level energy diagram of the transition in an atom, it is common to displace some of the levels horizontally to make the different transitions easier to visualize. For example, in the electronic energy levels of the hydrogen atom, shown in Figure 29.14, the pump must have energy that is greater than or equal to the difference between the ground state and the levels involved in the transition.

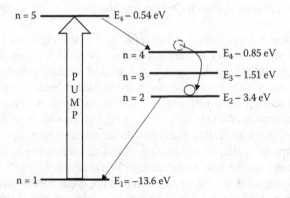

FIGURE 29.14 Laser level diagram of the 4→2 hydrogen transition.

In the case of a hydrogen atom in Figure 29.14, the pump is high voltage across the tube, which gives free electrons in the tube a kinetic energy that must be equal to or greater than the difference between the ground state and the fourth level, so that some electrons in hydrogen atoms are pushed into this higher level. Some electrons in the fifth level will move into the fourth level ($5^{th} \rightarrow 4^{th}$) and then some that are in the fourth level will transition down to the second level ($4^{th} \rightarrow 2^{nd}$), producing the photon described in the example. All the other transitions will also be present: $5^{th} \rightarrow 1^{st}$, $5^{th} \rightarrow 2^{nd}$, ... $4^{th} \rightarrow 3^{rd}$, $4^{th} \rightarrow 1^{st}$, $3^{rd} \rightarrow 2^{nd}$, ... and they will each produce photons, but in this example only the $4^{th} \rightarrow 2^{nd}$ transition occurs. Photons are absorbed and emitted by the energy levels presented in this discussion about hydrogen, but these energy levels are not laser levels since they do not have all the qualities required to make them suitable as a laser level. These qualities will be explained in the next section now that the structure of the diagrams has been established.

LASER LEVEL DIAGRAM

A laser level energy diagram of a four-level laser material, such as Nd:YAG, which is one of the most popular laser crystals, is shown in Figure 29.15. The horizontal displacement of levels 3 and 4 is only for visual convenience, as discussed in the previous section.

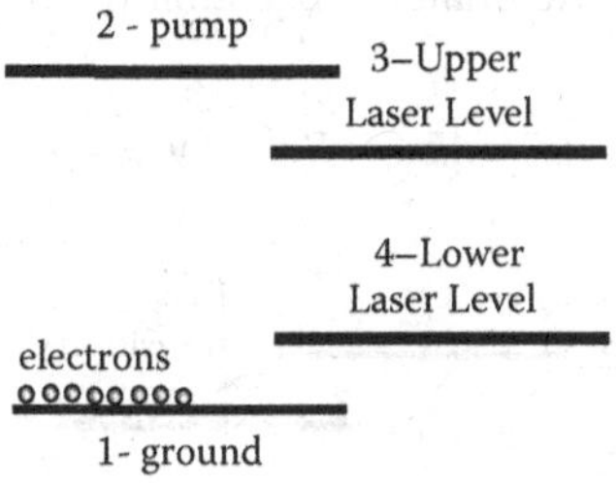

2 - pump

3–Upper
Laser Level

4–Lower
Laser Level

electrons

1- ground

FIGURE 29.15 A four-level laser diagram.

When the material is in its normal state, the electrons are in the ground state, as indicated by the circles in the ground state. It is important to note that the ground state is not the lowest level of the atom, but instead the level in which most of valance electrons reside when the atom is at standard temperature and pressure (STP) and not energized. The separation distance between the levels is an indication of the energy differences between the levels.

When the pump is turned on, the electrons are transferred from the ground level to an upper level, as depicted in Figure 29.16.

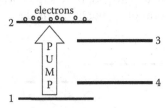

electrons

2

P
U
M
P

3

4

1

FIGURE 29.16 Electrons are moved to the pump level.

The energy difference from the ground state to the upper level determines the required energy per photon ($E = hv$) of the pump photons. In most atoms, the electrons will quickly transition back down the ground state, giving off energy in the form of photons on the way back down, either directly or through a series of steps, as described for the hydrogen atom. This is not the case for a substance that qualifies as a laser material.

The keys to a substance qualifying as a laser material are:

a. A fast and highly probable transition from the pump level 2 to the upper laser level 3. Most of the electrons in the pump level should move into upper laser level 3. The electrons that transition back down to the ground state are quickly pumped back up level 2 and then move over to the upper laser level.

b. The second property of the atomic structure is that there is a forbidden transition from upper laser level 3 to any other level in the atom. Since the transition from 3 to any other level occurs at a much slower rate than any of the other transitions in the material, this level 3 is called a metastable state. The reason that it is not called stable is that it does occur, but at a rate of about 1 million times less frequently than other transitions, so in practice it is stable but not completely.

With these material qualities and a pump source, electrons are trapped in the upper laser level, so the system looks like Figure 29.17, in which there are more electrons in an upper level than in the ground state. This condition is called *population inversion* and is a critical step in the possibility of a substance lasing. The wavelength of the laser light produced is determined by the energy level difference between levels 3 and 4. Rearranging equation (29.5) gives

$$\Delta E = |E_3 - E_4| = h\nu = \frac{hc}{\lambda}$$

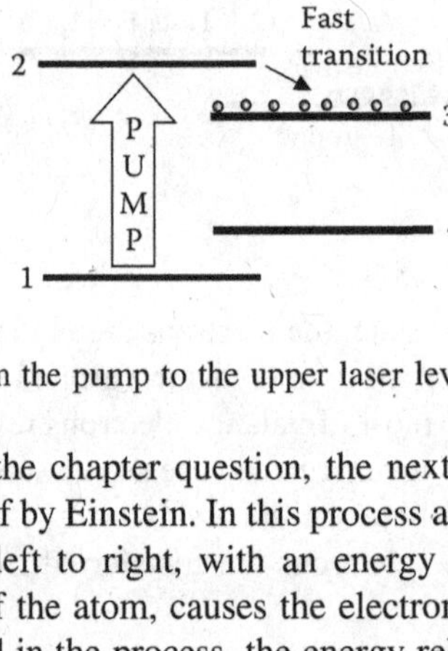

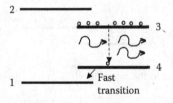

FIGURE 29.17   Fast transition from the pump to the upper laser level.

As mentioned in the answer to the chapter question, the next step in the process is **stimulated emission**, which was conceived of by Einstein. In this process a photon, the wave with the arrow in Figure 29.18 moving from the left to right, with an energy $(E_p = h\nu)$ that exactly equals the difference in the energy levels of the atom, causes the electron to transition from the upper laser level to the lower laser level, and in the process, the energy released by the electron transitioning from level 3 to level 4 is in the form of a photon that is the exact duplicate of the photon that initiated the transition.

FIGURE 29.18   The process of stimulated emission.

This is the process by which the laser light is produced. Since it depends on an atomic transition, the light is all one color, monochromatic, and since each photon that is made by this process is the exact duplicate of the one that created it, all the photons are in-phase, which is known as coherent light. These qualities of monochromaticity and coherence are what makes laser light special.

Since the upper and lower laser levels each have a slight spread of energy, the transitions can occur from the top of the level 3 to the bottom of level 4 or from the bottom of level 3 to the top of the level 4 or from anywhere on level 3 to anywhere on level 4. This gives a small spread in the possible energies and thus wavelengths of the monochromatic light produced by the laser. This small spread of wavelengths of the laser light is called the *bandwidth* of the laser. Most laser materials have a very small bandwidth, and there are many electrooptical techniques that can be used to reduce the bandwidth of lasers for specific applications. The coherence of the laser light is also a property that is initially established by the properties of the laser material and can be adjusted with other electrooptical techniques that are centered around the laser cavity.

Since the transition from level 4 to level 1 is highly probable, it is fast; therefore, the system quickly returns to its initial state until the pump is employed to start the process over again. For pulsed laser systems, the pump is turned on and off, sometimes very quickly, such as 5kHz (5000 cycles per second). Even though this is fast for us, it is not fast relative to the time in which the processes outlined in this section occur. So, there is plenty of time for the population inversion to build up and then deplete in the lasing process.

Example

Find the energy level difference of the laser levels in the Nd ion in Nd:YAG, given the laser wavelength of 1064 nm.

Solution

The wavelength of the photon produced by the laser transition in Nd:YAG is:

$$\lambda = 1064 \text{ nm.}$$

This is in the near infrared and corresponds to a photon frequency of:

$$\nu = \frac{c}{\lambda} = \frac{3 \times 10^8 \, m/s}{1064 \times 10^{-9} \, m} = 2.82 \times 10^{14} \, Hz$$

And an energy difference of:

$$\Delta E = h\nu = 6.626 \times 10^{-34} \, Js \cdot 2.82 \times 10^{14} \, Hz = 1.868 \times 10^{-19} \, J = 1.166 \, eV$$

THE CAVITY

Remember, Avogadro's Number is 6.022×10^{23} atoms/mole; thus, the number of laser atoms, even in a small laser crystal, is astronomical. So, even for a small piece of laser material, it is advantageous to make the light pass through the population inversion several times to cause as much amplification by stimulate emission as possible. To accomplish this feedback, mirrors are arranged to reflect the light back into the system. Since the light coming from the ends of the laser material diverges, as shown with the double-sided arrows extending from the ends of the laser material, it is often preferred to use mirrors with a curvature that matches the divergence of the light from the laser material, as depicted in Figure 29.19.

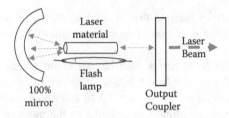

FIGURE 29.19 Laser cavity.

There are some phenomena, like thermal lensing, that need to be considered when designing a laser cavity, but as a start it can be understood as curved mirrors and lenses. Curved mirrors are positioned so that the light coming out the end of the laser material is reflected into the material. The photons generated by stimulated emission move in the same direction, so the mirrors guide the light back into the system, and the laser beam leaks out of the mirror that has less than 100% reflectivity, called an output coupler. This name comes from the fact that the laser light is coupled out of the system through it. The cavity creates the situation that results in a single direction laser beam.

Laser cavities are designed by ray tracing through the cavity and checking to see if the rays stay in the cavity. If the rays stay in the cavity, the cavity is stable. Many laser cavities employ two curved mirrors to improve the stability of the system.

When building a laser, there are often many adjustments that need to be made to the mirrors so the system will lase. If the mirrors are too far apart, much of the light leaving the laser material will not be reflected back into the system, as described in the Figure 29.20.

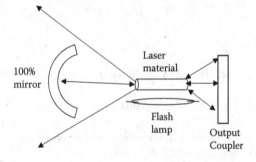

FIGURE 29.20 Unstable laser cavity.

This is called an *unstable cavity* and will most likely not lase, since there is too much loss in the system. Therefore, there will not be enough amplification by the stimulated emission. Even though there will be stimulated emission in the laser material, the net result will not be amplification, since there is more loss than amplification. The amount of amplification of a laser system is called *gain*. Thus, after population inversion, the next critical condition for lasing is that there must be more gain than loss.

In addition to the mirrors function in helping to achieve more gain than loss, the mirrors together play another critical role. That is, the laser cavity not only provides the necessary feedback, but it also helps control the wavelength of the light produced by the laser. It does this in a similar manner that a piano string generates the sound you hear or how the shape of an organ pipe controls the pitch of the sound made by the instrument. That is, standing waves of light are generated inside the laser cavity between the mirrors, and the harmonics of these standing waves affect the light coming out of the laser. The frequency spectrum of the laser material is determined by the spread in energy of the upper and lower laser levels in the material and the spread in frequency of the harmonics of the laser cavity. Thus, the properties of the laser material and the harmonics of the laser cavity produced the frequency of the laser light produced. Therefore, in a laser the light has particle properties as the photons are generated one by one through stimulated emission and wave properties as the standing waves in the cavity refine the frequency of the photons.

Together, the interaction between photons and atoms in the laser material and the harmonics of the laser cavity, produces light that is as close as we can get to monochromatic and coherent. These unique properties of the laser light give us the ability to send information through fiber optics, map the surface of the earth, or perform surgery.

Thermodynamics of a Laser

Because the pumping process must supply more energy than is needed per photon to initiate the laser process, there is a great deal of excess heat that must be extracted from the system to keep it from overheating. In most cases, chilled water flows through a conductive housing, made of aluminum or another metal, in which the laser material and flashlamps are located. This chiller is basically a pump and a refrigerator. Please see the chapter on thermodynamics for an explanation of how a refrigerator works.

Thermoelectric Cooler

In laser systems in which a fluid-based heat extraction is not possible, like small diode-based laser systems, ones deployed in harsh environments, or ones in which weight is an issue, like in space or on a drone, a different type of heat extraction is needed. In these cases, lasers utilize one of the most interesting devices that most people have never heard of, a thermoelectric cooler (TEC). These devices operate based on the Peltier effect, in which a current flowing through a circuit with two different conductors moves energy in the form of heat in one direction. If the junction is made of copper and bismuth wire the temperature of the junction rises when current passes from copper to bismuth, and the temperature of the junction falls when the current flows from bismuth to copper. Therefore, if the current is made to flow from bismuth to copper, the junction can be used to extract heat from a system. Today, most of these devices are made of two different semi-conductors, one positive (p-type) and one negative (n-type). The top and bottom of the TEC are made of a thermal conductive ceramic and, in between these two layers, are alternating p and n semiconductors. The n-types are connected to the positive (+) side of the power supply, and the p-type semiconductors are connected to the negative (−) side of the power supply. When a current is run through the TEC, the top becomes cooler than its surroundings, and the bottom becomes hotter, since heat is pushed from the top to the bottom of the device by the flow of current. This can be used to extract excess heat from a laser crystal or keep the pump diode lasers at the correct temperature for best pumping efficiency.

The excess heat from the pump is extracted from the laser by putting the cold side of the TEC in thermal contact with the laser material and the hot side on a heat sink, which may just be a larger piece of metal with a computer fan attached to the back side. The work (W) into the TEC is in the form of an electrical current (I) and a voltage (V). Since power (P) is work (energy) per time $\left[P = VI = \frac{\Delta W}{\Delta t} \right]$, the work done in a time t is just the product of voltage, current, and time, $W = (VI)t$.

From the chapter on thermodynamics, the coefficient of performance (COP) of a refrigeration system is like the efficiency of the system. Since the goal of a refrigeration system is to cool down an area, the heat extracted (Q_L) by the system is the task at hand and the work (W) that is needed accomplish that task make up the COP:

$$COP = \frac{Q_L}{W}$$

Since power is work per time, $P = W/t$ and $P = VI$ for an electrical device, the work done by a TEC is $(VI)t$, so the COP is:

$$COP = \frac{Q_L}{(VI) \cdot t}$$

From the study of these devices, the thermoelectric cooler is considered to be an irreversible Carnot-like reversed heat engine.[1] So, the COP of a TEC is

$$COP = \frac{T_L}{(T_H - T_L)}$$

Utilizing the two expressions of the COP and of the TEC, an expression of the rate of heat extracted (Q_L/t) by the TEC can be found as

$$\frac{Q_L}{t} = (VI) \frac{T_L}{T_H - T_L}.$$

So, given a specific voltage (V) of the TEC power supply, a specific rate of heat extracted (Q_L/t) can be accomplished for a known temperature difference T_L and T_H, by setting the current (I) of the TEC.

A closer look at the TEC process helps explains why it can be considered a Carnot-like heat engine. The hot and cold junctions are assumed to be isothermal, and they have negligible heat and electrical resistances compared with the resistances of the semiconductor materials n and p. Referring to the PV-diagram in Figure 29.21, the heat source at T_L transfers heat to the cold junction in step from D → C, the movement of the charge carriers from the cold to the hot is adiabatic from C → B, the heat exchanger at the hot junction at T_H takes place isothermally from B → A, and the final adiabatic transfer of charge carriers from the hot to the cold is from A → D. Of course, the flow of charge carriers is not perfectly adiabatic, and this is why it is called a Carnot-like heat engine operated in reverse.

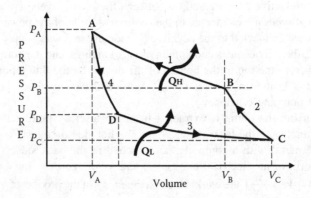

FIGURE 29.21   PV Diagram of the Carnot cycle run in reverse.

Example

Find the current needed to extract heat from a laser material at a rate of 3 W if the TEC if the system is set to 10 V and the temperature of the surroundings is 20°C and the cold sider of the TEC is 15°C. Also, find the COP of this TEC.

Solution

Solve the expression for current: $I = \left(\dfrac{1}{V}\right)\left[\dfrac{T_H - T_L}{T_L}\right]\dfrac{Q_L}{t}$

First, recognize the Q_L/t is the rate of heat extracted, which is power, so the unit is Watts (W)

$$\frac{Q_L}{t} = 3 \text{ W}$$

Convert the temperatures to kelvin:

$$T_H = 20°C + 273.15 = 293.15 \text{ K}$$
$$T_L = 15°C + 273.15 = 288.15 \text{ K}$$

Plug into the expression for I:

$$I = \left(\frac{1}{10V}\right)\left[\frac{293.15 \text{ K} - 288.15 \text{ K}}{288.15 \text{ K}}\right]3 \ W = 0.0052 \text{ A} = 5.2 \text{ mA}$$

The COP of this TEC is:

$$COP = \frac{T_L}{T_H - T_L} = \frac{288.15 \ K}{293.15 \ K - 288.15 \ K} = 57.63$$

1. El Haj Assad, M., "Thermodynamic Analysis of Thermoelectric Coolers," *International Journal of Mechanical Engineering Education*, 26, No. 3, 201–209, 1997.

HISTORICAL DEVELOPMENT OF THE LASER

The development of the laser has been well documented by many authors,[1] including accounts by some of the key figures in the creation of this important and ubiquitous technology.[2-4] These accounts highlight the path from Einstein's concept of simulated emission,[5] to Townes' maser,[6] to Townes and Schawlow's subsequent proposal for an optical maser,[7] and the eventual success of Maiman[8] in producing the first operating laser.

The first device to employ stimulated emission in the amplification of radiation was a MASER (Microwave Amplification by Stimulated Emission of Radiation) created by Gordon, Zeiger, and Townes[6] in 1953. This maser achieved stimulated emission by passing microwaves, generated in a klystron, through a cylindrical copper cavity filled with excited ammonia (NH_3) molecules. The excited ammonia molecules were separated from the molecules in the lower state by an external electric field before they were injected into the cavity. In 1958, Schawlow and Townes[7] proposed the development of infrared and optical masers, which opened the doors to the research and development of the laser. The eventual winner of the technological race to build the first laser was won by T. Maiman[8] in 1959. It is clear that Schawlow and Townes' paper stimulated the research efforts that eventually lead to Maiman's Ruby laser and, it is also clear, that Schawlow and Townes were careful to site the work of N. Bloembergen[9] on solid state masers in their work. In his 1956 paper, Bloembergen suggested that structure of the energy levels in a solid could be used to store the excited molecules and thus separate them from the lower level molecules to create a population inversion, where more of the molecules in the crystal are in an excited state than in the ground state. This proposal was not only important to Maiman's success, but also to the entire modern laser industry. This is the case, because, unlike in the first maser, which actively separated out the excited molecules, the energy-level structure of a laser material is used to store energy for the eventual stimulated emission.

The first operation of a Solid-State Maser, which was proposed by Bloembergen,[9] was achieved by Scovil, Feher, and Seidel[10] in 1956. In their paper, the authors report on achieving success in implementing the proposal of Bloembergen using magnetically diluted paramagnetic salt. This system consisted of a paramagnetic salt in which the Gd++ ion is pumped into excited state with a magnetic field and amplification is achieved for the microwave radiation at 9 GHz ($\lambda = 3.33$ cm). This achievement is the first account of a solid-state, stimulated-emission amplification system that did not rely on an external separation of the excited molecules, but instead, utilized the energy-level structure to store the energy eventually used for amplification by stimulated emission.

The other important discovery that may have had a direct impact on Maiman's choice of material, was the work of Makhov, Kikuchi, Lambe, and Terhune. In two papers,[11,12] these authors report on masers developed using Ruby (Al_2O_3:Cr) crystals. The authors make a point of explaining the role of the trivalent chromium ion, Cr^{+++}, as the maser ion in the crystal. This is suggestive of the language used today for common laser material like Nd:YAG, where the Neodymium (Nd^{+++}) is the laser ion in the crystal of Yttrium Aluminum Garnet ($Y_3Al_5O_{12}$). In the papers, the authors explain that forbidden transitions in the chromium ions are pumped via specific transitions with a magnetic field. The crystal is oriented in a cylindrical cavity designed to excite specific modes of oscillation. This language is also suggestive of that which is used in a modern paper on the development of a new laser material.

BIBLIOGRAPHY FOR THE HISTORICAL DEVELOPMENT OF THE LASER

1. Hecht, J., *Beam: The Race to Make the Laser*, Oxford University Press, NY, 2005.
2. Townes C., *How the Laser Happened: Adventures of a Scientist*, Oxford University Press, NY, 1999.
3. Maiman, T, *The Laser Odyssey,* Laser Press, Blaine, WA, 2000.
4. Taylor, N., *Laser: The Inventor, the Nobel Laureate, and the Thirty-Year Patent War,* Simon & Schuster, NY, 2000.
5. Einstein, A, "On the Quantum Theory of Radiation", *Phys. Z.,* 18, 121, 1917.
6. Gordon, J., Zeiger, H., and Townes, C., "Molecular Microwave Oscillator and New Hyperfine Structure in the Microwave Spectrum of NH3", *Physical Review,* 95, 282–284, July, 1954.

APPENDIX QUESTIONS AND PROBLEMS

The first laser to operate successfully, in 1959, was the ruby laser. The basic structure of the laser is a ruby crystal with one side coated with a 100% reflective and a partially reflective surface on the right. A flashlamp was wrapped around the ruby crystal, and the lamp was powered by a high voltage supply.

Problem 1

E_1 is the ground state of the chromium ions, and energy is −4.02 eV. E_{2a} and E_{2b} are the pump band for the chromium ions. E_3 is the metastable laser level in chromium ions. It has two distinct levels, as seen in Figure 29.22. The electrons in the ground state are moved to the pump bands by pump photons from a flashlamp. The ions quickly move from the pump bands to the laser levels, where they are trapped in the metastable states. The transition from the upper laser levels to the ground state is initiated by other photons that initiate stimulated emission that brings the ions back to the ground state and results in the amplification of the light. Compute the energy in eV of the upper laser level (E_3) from which a transition directly down to the ground state results in the laser photon with a wavelength of 694.3 nm. The photon from the lower laser level of (E_3) is at 692.7 nm.

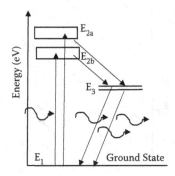

FIGURE 29.22   Ruby crystal laser level diagram.

MULTIPLE CHOICE QUESTIONS

1. Is the energy of each of the 694.3 nm laser photons greater than, less than, or equal to the energy of each of the 692.7 nm laser photons?
   A. greater than
   B. less than
   C. equal to

2. Are the wavelengths of the laser photons longer than, shorter than, or the same as the wavelengths as the pump photons?
   A. longer than
   B. shorter than
   C. the same as

3. Must the frequencies of the pump photons needed to move electrons in the Cr3+ ions from the ground state to the E_{2a} energy level have a higher, a lower, or the same frequency as the pump photons needed to move the electrons in the Cr3+ ions from the ground state the E_{2b} level?
   A. higher
   B. lower
   C. the same

4. For this system to lase, population inversion must be achieved. This occurs when there are more electrons in which level, E_{2a}, E_{2b}, or E_3, as compared to the number of electrons in the ground state?
   A. E_{2a}
   B. E_{2b}
   C. E_3

5. Stimulated emission occurs in a transition between which two atomic states?
   A. $E_1 \rightarrow E_{2a}$
   B. $E_1 \rightarrow E_{2b}$
   C. $E_{2a} \rightarrow E_3$
   D. D. $E_{2b} \rightarrow E_3$
   E. E. $E_3 \rightarrow E_1$

Problem 2

The laser system that is responsible for producing most of the light that moves through the fiber optics that makes up much of our communication system is an erbium-doped fiber laser. Figure 29.23 shows the energy level diagram of the erbium ions (Er^{3+}) that are mixed into the glass to make the fiber laser. The energy of the Er^{3+} ion levels are given in eV. Compute the wavelength of the laser photons produced.

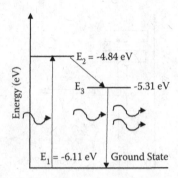

FIGURE 29.23 Erbium ion laser level diagram.

MULTIPLE CHOICE QUESTIONS

6. Is the energy of each of the laser photons greater than, less than, or equal to the energy of each of the pump photons?
    A. greater than
    B. less than
    C. equal to

7. Is the wavelength of the laser photons longer than, shorter than, or the same as the wavelength of the pump photons?
    A. longer than
    B. shorter than
    C. the same as

8. To pump electrons in the Er^{3+} ions from the ground state to a pump band with an energy of -5.57 eV as compared to the pump band in the diagram with an energy of -4.84 eV, would pump photons require longer, shorter, or the same wavelength?
    A. longer
    B. shorter
    C. the same

9. For this system to lase, population inversion must be achieved. This occurs when there are fewer electrons in the ground state than in which level?
    A. E_1
    B. E_2
    C. E_3

10. Stimulated emission occurs in a transition between which two atomic states?
    A. $E_1 \rightarrow E_2$
    B. $E_2 \rightarrow E_3$
    C. $E_3 \rightarrow E_1$

30 Nuclear Structure & Radioactivity

30.1 INTRODUCTION

In this volume of the textbook, energy has played a central role in many of the calculations and discussions, so it is fitting to end the textbook with a topic that requires understanding of one of the most famous equations in all of science, $E = mc^2$. In the first part of the chapter, the structure of the nucleus and the energy required to keep the nucleus together are presented. The mass difference between the individual particles that make up the nucleus and that of the nucleus itself is found, and the binding energy of the nucleus is computed using Einstein's famous equation. Examples of the mass difference associated with nuclear processes, such as fission and fusion, are presented, and calculations of each process are provided. As an extension of this analysis, the energy released in another natural process, radioactivity, is discussed, and more examples are provided. In addition, the transmutation of elements due to radioactive processes is addressed, and the time dependence of these processes is studied with the concept of half-life. Several examples applying the exponential expressions associated with radioactive half-life and radioactive decay constants are provided. By the end of the chapter, you should feel comfortable applying Einstein's expression, $E = mc^2$, and explaining how energy and mass are exchanged at the nuclear level.

30.2 CHAPTER QUESTION

To kill bacteria, some food is exposed to radiation. Does this make the food radioactive? To answer this question requires an understanding the structure of the nucleus and of radioactivity. These topics will be covered in this chapter, and this question will be answered at the end of the chapter.

30.3 THE NUCLEUS

The nucleus of an atom is made up of protons and neutrons. The proton has a charge of 1e and a mass about 2000 times the mass of an electron. A neutron is electrically neutral and has a mass that is a little bit larger than that of the proton. When protons and neutrons are in the nucleus of an atom, they are referred to collectively as *nucleons*.

The number of protons, denoted by the atomic number, determines the element. For example, the element depicted in Figure 30.1 is helium (He) since it has two protons in the nucleus.

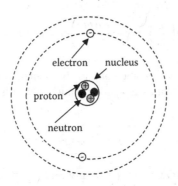

FIGURE 30.1   Helium atom.

DOI: 10.1201/9781003308072-31

In this depiction, the helium nucleus also has two neutrons in the nucleus. The total number of nucleons, which is the atomic mass number, is four for this atom. An isotope of an element is an atom of that element with a particular number of neutrons. Different isotopes of the same element have different numbers of neutrons, but the same number of protons. In the notation used to represent an isotope, the atomic number is commonly written as a subscript before the atomic symbol, and the atomic mass number is written as a superscript before the atomic symbol, so the isotope of helium depicted in Figure 30.1 is written as $\left(^{4}_{2}\text{He}\right)$.

The atomic number is often omitted as it is implied by the name of the element. Hence, the depicted isotope of helium is often designated as ^{4}He. Another notation, referred to as *hyphen notation*, involves writing the full name of the element followed by a hyphen and then the atomic mass number. In this notation, the depicted isotope of helium is designated helium-4.

On the periodic table of elements, a weighted average of the masses of the isotopes of an element is given as the mass of an atom of that element. In calculating that average, the weighting factors are the relative abundances of the various isotopes of the element found in nature. The unit of mass is the atomic mass unit u, which is defined to be one-twelfth the mass of a carbon-12 atom,

$$1 \text{ u} = 1.66054 \times 10^{-27} \text{ kg}$$

30.3.1 BINDING ENERGY ($E = mc^2$)

An interesting fact about atoms is that the mass of an atom is less than the sum of the masses of its parts. The missing mass is converted into the *binding energy*, which binds the nucleus together. This is a simple statement that is difficult to accept, so some additional discussion is required. All individual protons, when found outside a nucleus, have the same mass. This is also the case for all individual neutrons. On the other hand, when these particles are together in a nucleus, their mass changes. In fact, all nuclei are made up of protons and neutrons, but the mass of a nucleus is always less than the sum of the individual masses of the protons and neutrons of which it is made. It is clear that the physics of the nucleus is different than that of our daily experience. It is ridiculous to think that the mass of a dozen eggs in a carton is less than the sum of the mass of 12 individual eggs, but when nucleons are in a nucleus, they have less mass then when they are outside the nucleus. This concept is formalized by a very famous equation, given in equation (30.1) as

$$E = mc^2 \tag{30.1}$$

where E is energy, m is mass, and c is the speed of light. Since in the SI system, the units of mass are kg and the units of the speed of light are $\left(\frac{m}{s}\right)$, so the combination of these SI units into mc^2 results in Joules (J),

$$kg\left(\frac{m}{s}\right)^2 = \left(\frac{(kg)(m)}{s^2}\right)m = (N)m = J$$

When referring to the combination of nucleons in the nucleus of an atom, the m in mc^2 is the missing mass and thus often written as Δm. So, equation (30.1) is commonly applied in the form given in equation (30.2) as

$$E = (\Delta m)c^2 \tag{30.2}$$

Remember that protons are positive, so they strongly repel each other in the close confines of the nucleus. The missing mass provides the energy required to overcome the electrical potential energy of the protons repelling each other. Because this energy results in the nuclear strong force, this energy is referred to as the ***binding energy*** of the nucleus.

Example 1

An alpha particle, which is a Helium nucleus shown in Figure 30.2, is made up of two protons and two neutrons.

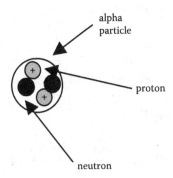

FIGURE 30.2   Alpha particle.

The mass of a separate proton is 1.00728 u, and the mass of each separate neutron is 1.00866 u. Adding the mass of 2 protons and 2 neutrons gives:

$$[2(1.00728 \text{ u}) + 2(1.00866 \text{ u})] = 4.03188 \text{ u}.$$

The mass of an alpha particle is 4.00153 u.

This results in a missing mass (Δm) of: $\Delta m = 4.03188 \text{ } u - 4.00153 \text{ } u = 0.0304 \text{ } u$, which results in a binding energy of

$$E = (\Delta m)c^2 = (0.0304 \text{ } u)\left(1.66053886 \times 10^{-27} \tfrac{kg}{u}\right)\left(3 \times 10^8 \tfrac{m}{s}\right)^2$$
$$E = 4.5565 \times 10^{-12} \text{ } J$$
$$E = (4.5565 \times 10^{-12} \text{ } J)\left(\tfrac{1 \text{ } eV}{1.602 \times 10^{-19} \text{ } J}\right) = 28,442,571.8 \text{ } eV$$

Although this may seem like a small number, it is approximately 500,000 times greater than the energy that binds the electrons to atoms. Remember that the ionization energy of Hydrogen is 13.6 eV, and the ionization of Helium is

$$E_n = 13.6 \text{ eV } (Z^2) = 13.6 \text{ eV } (2^2) = 54.4 \text{ eV}$$

as compared to the binding energy of a Helium nucleus, which is an alpha particle (He^{++}), which is 28,442,571.8 eV. Thus, any process that deals with the binding energy is significantly more powerful than electrical or chemical phenomenon.

Example 2

Consider for instance, the mass of a carbon-12 atom, which has a mass of exactly 12 u, because carbon-12 was chosen as the reference isotope for the unit u. The carbon-12 atom consists of 6 protons, 6 neutrons, and 6 electrons. To avoid having to treat the electrons separately, the carbon-12 atom can be considered to be comprised of 6 neutrons and 6 hydrogen atoms. The mass of a neutron is 1.008664 u, and the mass of the hydrogen atom ^1H is 1.007825 u. Hence, the sum of the masses of the parts of a carbon atom is:

$$m_{parts} = 6m_n + 6m_{Hydrogen-1}$$
$$m_{parts} = 6(1.008665 \ u) + 6(1.007825 \ u)$$
$$m_{parts} = 12.09894 \ u$$

Therefore, the missing mass for carbon-12 is:

$$\Delta m = m_{parts} - m_{C-12} = 12.09894 \ u - 12 \ u = .09894 \ u$$

Convert this missing mass from u to kg:

$$\Delta m = .09894 \ u = .09894 \ u \left(\frac{1.66054 \times 10^{-27} kg}{u} \right) = 1.64294 \times 10^{-28} \ kg$$

Next, calculate the binding energy and convert it to eV.

$$E_B = (\Delta m)c^2 = (1.64294 \times 10^{-28} \ kg) (2.9979 \times 10^8 \ m/s)^2 = 1.4766 \times 10^{-11} \ J$$
$$E_B = (1.4766 \times 10^{-11} \ J) \frac{1eV}{(1.6022 \times 10^{-19} C)J/C} = 9.2161 \times 10^7 \ eV$$

The mass deficit times c^2 is the minimum amount of energy that one would have to add to the atom to break it up into its constituent parts. A quantity that is useful for comparing different nuclei is obtained by dividing the binding energy (E_B) of a particular nucleus by the number of nucleons, #N, in that nucleus. For the case of carbon-12, the binding energy per nucleon in eV is:

$$\frac{E_B}{\#N} = \frac{9.2161 \times 10^7 \ eV}{12} = 7.6801 \times 10^6 \ eV = 7.6801 \ MeV$$

The prefix M is mega and stands for 1×10^6 or one-million. Compare this with the binding energy per electron in the hydrogen atom (13.6 eV). The binding energy per nucleon is over 500,000 times the binding energy of the electron in hydrogen.

30.3.2 Binding Energy Curve

As demonstrated in the previous example, the binding energy of the nuclei is found by multiplying the missing mass (Δm) of each element by c^2 and then dividing the binding energy by the number of nucleons (protons + neutrons). A graph of the binding energy per nucleon as a function of the atomic mass, the number of nucleons in the nucleus of the atom, for the atoms in the periodic table is given in Figure 30.3.

Since the nucleus of the isotope of hydrogen that consists of one proton $\left(^1_1H\right)$ the binding energy of this nucleus is zero. From the graph, it is apparent the binding energy per nucleon increases rapidly from hydrogen to helium and then approaches a maximum near iron and nickel. The isotope with the most binding energy per nucleon is nickel-62, $\left(^{62}_{28}Ni\right)$ at 8.7948 MeV.

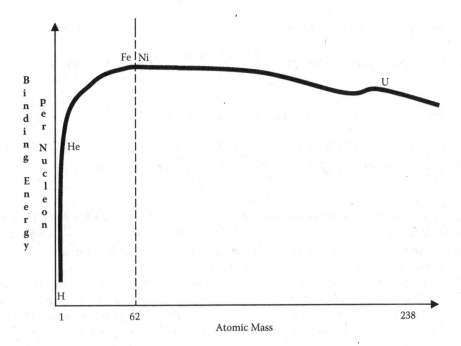

FIGURE 30.3 The binding energy curve.

30.3.3 NUCLEAR FUSION

If two low-mass elements combine to form an element having a mass that is less than or equal to the mass of iron, the nucleons in the new element are more strongly bound than they were before combining. Being more strongly bound means that an atom of the new element has less mass than the combined mass of the two original atoms. This means that some of the mass of the original two atoms was converted into energy. This process is known as *nuclear fusion,* and it is the source of energy that powers stars, including our sun.

For example, part of the fusion process that powers the sun includes an interaction that combines two nuclei of an isotope of hydrogen called deuterium $\left(^2_1H^+\right)$ to form a helium nucleus, also known as an alpha particle, $\left(^4_2He^{2+}\right)$ with release of energy in the form of a gamma ray photon (γ). The single + in the notation for the deuterium ion means it is missing one electron and hence is an ion with charge +1e. The "2+" in the notation for the helium ion means it is missing both of its electrons, leaving it with a charge of +2e. This process is given in the following nuclear equation,

$$^2_1H^+ + ^2_1H^+ \rightarrow ^4_2He^{2+} + \gamma$$

The energy released in the process can be found by finding the missing mass difference from before to after the process and using equation (30.2).

Starting with the mass of each of the particles involved in the process:

$$m_{h2} = 2.01355321270 \ u \quad m_{he4} = 4.001506179 \ u$$

Therefore, the mass difference is the mass before, in the form of the sum of the mass of the two isotopes of hydrogen, and the mass after, in the form of the helium isotope. The gamma (γ) particle does not have a mass, since it is a form of e&m wave.

$$\Delta m = (m_{h2} + m_{h2}) - m_{he4} = 2(2.01355321270 \ u) - 4.001506179 \ u = .0256002464 \ u$$

Convert this missing mass from u to kg:

$$\Delta m = .0256002464 \ u = .0256002464 \ u\left(\frac{1.66054 \times 10^{-27} \ \text{kg}}{\text{u}}\right) = 4.25102332 \times 10^{-29} \ \text{kg}$$

Next, calculate the energy released in the process and convert it to eV.

$$E = \Delta mc^2 = (4.25102332 \times 10^{-29} \ \text{kg}) \ (2.9979 \times 10^8 \ \text{m/s})^2 = 3.82056657 \times 10^{-12} \ \text{J}$$

$$E = (3.82056657 \times 10^{-12} \ \text{J})\frac{1\text{eV}}{(1.6022 \times 10^{-19}\text{C})J/C} = 2.3845753 \times 10^7 \ \text{eV}$$

This process occurs in the core of the sun and results in the release of 23.846 MeV per combination. Since the process occurs with four nucleons, there is a release of $\frac{23.846 \ MeV}{4} = 5.9615$ MeV per nucleon. In the sun, these gamma rays heat the sun from the center out to the surface of the sun. Of course, two hydrogen nuclei, which are both positive, will repel each other. Thus, for this process to occur, these particles must be moving so fast that even the strong electrostatic repulsion will not be enough to stop them from colliding and combining. Therefore, the temperature of the core of the sun must be approximately 15,000,000 K for this to occur. This is one of the reasons that controlled nuclear fusion, for purposes of an energy source, is so difficult to achieve here on earth.

The creation of elements in the nuclear fusion processes in stars is limited to elements with an atomic number less than iron, since the fusion of iron would require energy rather than provide it. This is the first part of the binding energy curve, Figure 30.3, from hydrogen, H, up to nickel, Ni, since the curve increase rapidly then levels off and reaches a peak. This explains why elements with atomic masses less than iron are the most common and are the building blocks of most of the matter. Fusion reactions in which the end product has an atomic mass less than or equal to that of iron can be self-sustaining, in that energy released in the fusion of two atoms can be used in the fusion of other atoms. Fusion processes involving an end product with an atomic mass greater than that of iron are less energetically favorable, and in fact, in most cases, use energy rather than release it. Such elements are formed in the explosions of high mass stars known as supernovas.

30.3.4 NUCLEAR FISSION

On the other side of the binding energy curve from nickel (Ni) out to elements like uranium (U), the heavy elements must be broken apart to produce other elements with less atomic mass to release energy. If the atomic mass of at least one of the resulting elements is still greater than that of iron, the process can release energy. This process is called *nuclear fission*.

For example, the fission process that is used for some nuclear power and nuclear weapons is the fission of uranium-235 ($_{92}U^{235}$). This process is initiated by firing a low energy neutron (n) into a nucleus of uranium-235. This collision splits the uranium-235 into the nuclei of two other elements, barium ($_{56}Ba^{141}$) and krypton ($_{36}Kr^{92}$), and releases three neutrons and kinetic energy in the form of the moving particles.

$$^{235}_{92}U + n \rightarrow ^{141}_{56}Ba + ^{92}_{36}Kr + 3n + \text{Energy}$$

The energy released in the process can be found by finding the missing mass difference from before to after the process and using equation (30.2).

Starting with the mass of each of the particles involved in the process:

$$m_{U235} = 235.043923 \ u \quad m_n = 1.008665 \ u$$
$$m_{Ba141} = 140.91440 \ u \quad m_{Kr92} = 91.92630 \ u$$

Therefore, the mass difference is the mass before, in the form of the sum of the mass of the isotope of Uranium and the neutron, and the mass after, in the form of the isotopes of Barium and Krypton and the three neutrons.

$$\Delta m = (m_{U235} + m_n) - (m_{Ba141} + m_{Kr92} + 3 \ m_n)$$
$$\Delta m = (235.043923 \ u + 1.008665 \ u) - [(140.91440 \ u) + (91.92630 \ u) + 3(1.008665 \ u)]$$
$$\Delta m = 0.185893 \ u$$

Convert this missing mass from u to kg:

$$\Delta m = 0.185893 \ u = 0.185893 \ u \left(\frac{1.66054 \times 10^{-27} \ kg}{u} \right) = 3.08682762 \times 10^{-28} \ kg$$

Next, calculate the energy released in the process and convert it to eV.

$$E = \Delta mc^2 = (3.08682762 \times 10^{-28} \ kg)(2.9979 \times 10^8 \ m/s)^2 = 2.77425682 \times 10^{-11} \ J$$
$$E = (2.77425682 \times 10^{-11} \ J)\left(\frac{1eV}{(1.6022 \times 10^{-19}C)J/C} \right) = 1.731529658 \times 10^8 \ eV$$

Which is approximately 173.2 MeV per fission process and $\frac{173.2 \ MeV}{236} = 0.7339$ MeV per nucleon. There is substantially less energy per nucleon than in the example of the fusion reaction of the sun. This difference is consistent with the steep slope of the binding energy curve, Figure 30.3, from hydrogen up to iron as compared with the gradual slope of the curve from uranium back towards nickel. In addition, the form of the energy is different in the fission process and the fusion processes in the examples. The fusion example that happens as part of the sun's processes results in energy in the form of gamma rays that heat the sun. In this fission of uranium, most of the energy is released in the form of kinetic energy of the resulting particles, known as daughter nuclei, along with the positron and the neutrino. In addition, in the fission process, each of the three neutrons given off can cause the nucleus of another uranium atom to split apart, which releases more neutrons, which cause more fission events, and so on. The process is referred to as a chain reaction and is the key to catastrophic uncontrolled reaction that occurs in nuclear weapons. It turns out that one of the biggest challenges to producing nuclear weapons is separating out the rare isotope of uranium-235, which releases these extra neutrons in its fusion reaction, from the much more common uranium-238 that does not.

30.4 RADIOACTIVITY

Radioactivity is a natural process in which the nucleus of an atom releases energy in the form of ejected particles and/or e&m waves (radiation), that was discovered in 1896 by Henri Becquerel. In 1898, Marie and Pierre Currie were the first to isolate a radioactive element, which was radium, that has 88 protons in its nucleus. The most common isotope of radium found on earth has 138 neutrons in the nucleus, bringing the total number of nucleons (protons and neutrons) to 226. Thus, the most common isotope of radium is radium-226 $\left(^{226}_{88}Ra \right)$. This radioactive isotope naturally decays into radon-222 $\left(^{222}_{86}Rn \right)$. In the process, it loses two protons from having 88 protons to having 86 protons, and it loses two neutrons from having a total of 226 nucleons to having 222 nucleons. The particle ejected from the nucleus of radium is a helium nucleus $\left(^{4}_{2}He^{2+}\right)$, referred to in this context as an alpha (α) particle. This process of changing from one element to another is known as *transmutation* and is often expressed in the form of a reaction equation such as:

$$^{226}_{88}Ra \rightarrow\ ^{222}_{86}Rn + \alpha$$

where energy is released in the form of kinetic energy of the products. As more experiments were conducted with other radioactive elements, more types of particles were discovered.

The three most common types of radioactivity emit either alpha (α) or beta (β) particles or gamma (γ) radiation. As mentioned above, the alpha particle (α) is a helium nucleus $\left(^{4}_{2}He^{2+}\right)$. A beta particle ($\beta$) is an electron ($e^-$) in the case of β^- decay, and positron (e^+) in the case of β^+ decay. A positron is the antiparticle of an electron, so it has the same mass as an electron, but it has a charge $+1e$ rather than $-1e$. When the algebraic sign is not specified, the beta decay under discussion is β^- decay. Whenever an electron is emitted (β^- decay), an electron antineutrino $\bar{n}_e$ is also emitted, and whenever a positron is emitted (β^+ decay), an electron neutrino v_e is also emitted. Neutrinos are uncharged, very low mass particles that are very difficult to detect. Gamma (γ) rays are photons having the highest range of frequencies of all electromagnetic radiation. Alpha decay and beta decay represent transmutations from one element to another, but in gamma decay, the isotope retains its original identity.

The units for the decay rate are the same for all three processes. The SI unit of decay rate is the becquerel (Bq), which is one decay per second. So, the decay rate is often listed in units of just decays per second, which is abbreviated s^{-1}. The curie (Ci) is 3.7×10^{10} decays per second, which is approximately the decay rate of 1 g of radium.

30.4.1 RADIOACTIVE HALF-LIFE

The half-life, which is denoted with the symbol ($t_{1/2}$), for a given radioactive isotope of an element is the time it takes for half the radioactive nuclei in any sample of the isotope to undergo radioactive decay. No matter the number of particles the sample starts with, or when the counting starts, the half-life is always the same. After one half-life, there will be one-half of the radioactive nuclei remaining in the sample from when the count was started. Remember, the particles that undergo radioactive decay don't disappear – they just change from one element to another in a transmutation process. Since the half-life is constant, the probability that any nuclei at any given time is the same for any other of the same type of radioactive nuclei in the sample. So, we don't know when any specific nuclei will undergo a radioactive process, but over the time of one half-life, we know that half of the nuclei in a sample will undergo a radioactive decay.

That is, the half-life for a given radioactive isotope is based on the rate of decay of that isotope. Since the time for one-half the radioactive substance to decay is consistent, no matter how much or how little of the radioactive substance is in the sample, the rate of decay must be proportional to the quantity of the radioactive material in the sample. So, when there is a large number of radioactive particles in the sample, more radioactive decays occur per second, and when there are very few radioactive decays, there are only a few decays per second. Thus, the rate of change of the decay rate is proportional to the quantity of the radioactive material present in the sample under consideration. As with the study of the discharging capacitor, the mathematical function that represents a decrease in the number of radioactive elements (N) in as sample is given in equation (30.3) as

$$\frac{\Delta N}{\Delta t} = -N\lambda \tag{30.3}$$

where λ is the decay constant of the radioactive element. This lambda (λ) has nothing to do with wavelength, but is the common symbol used for the decay rate. The solution to equation (30.3) is given in equation (30.4) as

$$N = N_o e^{-\lambda t} \qquad (30.4)$$

This expression provides a way to find the number of radioactive elements (N) in a sample at a given moment of time (t), if the number of radioactive elements at $t = 0$ s, (N_o), is known.

A simple example of a graph of the number of atoms of a radioactive isotope remaining vs. time is shown in Figure 30.4.

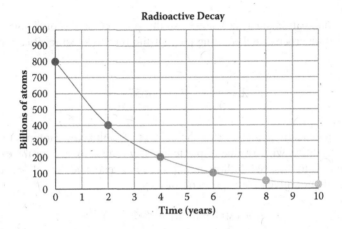

Radioactive Decay

FIGURE 30.4 Graph of number of radioactive atoms remaining in a sample as a function of time.

In this example, a sample starts with 800 billion radioactive atoms that have a radioactive half-life of 2 years. So, in 2 years, 400 billion of these radioactive isotopes decay and 400 billion remain. In the next 2 years, half of the remaining 400 billion, or 200 billion, decay and so on. Thus, a constant *percentage* of the remaining atoms decays in a specified amount of time and not a constant *number* of atoms. This is the defining characteristic of an exponential decay.

Using the half-life of 2 years from the example depicted in Figure 30.4 and the preceding paragraph, the value of the decay constant (λ) for this case can be found and then used to compute the number of radioactive atoms remaining at any time.

First, to find the decay constant, divide both sides of equation (30.4) by N_0 to get:

$$\frac{N}{N_0} = e^{-\lambda t}$$

Next, take the natural log of both sides to get:

$$\ln\left(\frac{N}{N_0}\right) = -\lambda t$$

Solve for λ by multiplying both sided by (-1) and dividing by the time (t), which results in equation (30.5) as

$$\lambda = -\frac{1}{t} \ln\left(\frac{N}{N_0}\right) \qquad (30.5)$$

Equation (30.5) is a solution of equation (30.4) for the decay constant, so it is not actually a new expression, but since it provides the expression that is required to calibrate the decay constant for any specific radioactive processes, it has been acknowledged as a separate equation.

So, to calibrate the decay constant for any radioactive process, substitute the half-life ($t_{1/2}$) for the time (t) in the equation (30.5) and the value of $\left(\frac{1}{2}\right)$ for the ratio of $\left(\frac{N}{N_0}\right)$, since the half-life is the time for the N to reach one-half of N_o. Making these substitutions into equation (30.5) results in

$$\lambda = -\frac{1}{t_{1/2}} \ln\left(\frac{1}{2}\right) = -\frac{1}{t_{1/2}}(-0.693) = \frac{0.693}{t_{1/2}}$$

The natural log of one-half is always (-0.693), so this previous step should always have the same result in the calibration process. Substituting the half-life for this specific situation of $t_{1/2}= 2$ years, results in

$$\lambda = \frac{0.693}{2 \ years} = .34657\frac{1}{years}$$

Equation (30.4) can be calibrated to predict the number of radioactive isotopes that are remaining at any given time t starting with a value of particles N_0 at time zero. So, for the example shown in Figure 30.4, starting with $N_0 = 800$ billion particles at $t = 0$ years and the calibrated decay constant from the previous section gives the expression of

$$N = (800 \ billion) \ e^{-(0.34657 \ years^{-1})t}$$

To find the number of radioactive particles remaining at 3 years, a value of $t = 3$ years in the previous expression, and the following value can be computed as

$$N = (800 \ billion) \ e^{-(0.34657 \ years^{-1}) \ 3yrs} = 282.85 \ billion.$$

Studying Figure 30.4, this seems to agree well with the value from the graph.

30.4.2 Combining Decay Constants

If a sample consists of atoms, each of which can decay two ways with two different decay rates, λ_1 and λ_2, the total activity decay rate (λ) will simply be the sum of the two individual decay rates,

$$\lambda = \lambda_1 + \lambda_2.$$

For example, an excited state of a silver nucleus designated ^{120*}Ag, where the * is the symbol used to indicate that the nucleus is in an excited state, can decay to the ground state 120Ag with the emission of a gamma ray (photon), or, it can decay to ^{119}Cd with the emission of an electron, through a beta decay. If, at some instant, in a sample containing ^{120*}Ag atoms, the rate of gamma decay is calibrated to a decay constant value of λ_γ and the rate of beta decay is calibrated to a decay constant value of λ_β, the total decay constant for the overall decay is given by

$$\lambda = \lambda_\gamma + \lambda_\beta$$

30.4.3 NUCLEAR DECAY AND RADIOACTIVE SERIES

30.4.3.1 Alpha Decay

As mentioned in the previous section, an alpha decay results in the ejection of a helium nucleus $\left(^4_2\text{He}^{2+}\right)$, which decreases the atomic number by 2 and the atomic mass number by 4. An example of an alpha emission is that of radon-222, which decays to polonium-218 with a half-life of approximately 3.8 days.

$$^{222}_{86}\text{Rn} \rightarrow\ ^{218}_{84}\text{Po} + \alpha$$

30.4.3.2 Beta Decay

β^- decay results in the ejection of an electron and an electron antineutrino from the nucleus. For this to happen, one of the neutrons in the nucleus must decay into a proton plus an electron and an antineutrino. This causes an increase in the atomic number of the nucleus of one, and since the atomic mass number gives the total of the number of protons + neutrons, there is no change in the atomic mass number of the element due to a beta decay. An example of a beta decay is that of cesium-137 into an excited state of barium-137:

$$^{137}_{55}\text{Cs} \rightarrow\ ^{137}_{56}\text{Ba}^* + e^- + \bar{\nu}_e$$

(The symbol for a nucleus in an excited state is marked with an asterisk as a right superscript on the chemical symbol for the element.)

30.4.3.3 Gamma Decay

Since gamma rays have no charge or mass, a gamma decay results in no change of the isotope's atomic number or atomic mass number. The only change is a decrease in the energy of the nucleus. Gamma emission occurs when the nucleon that is not in the lowest energy level available to it in an excited nucleus, jumps down to a lower energy level with the emission of a photon whose energy is the difference between the two energies. For example, after cesium-137 decays to an excited state of barium-137, the barium nucleus gives off a gamma ray in a process represented as:

$$^{137}_{56}\text{Ba}^* \rightarrow\ ^{137}_{56}\text{Ba} + \gamma$$

In many situations, a sequence of decays is common. In many cases, the sequence is quite extensive. For example, consider decay of neptunium-237 all the way down to bismuth-209:

$$
\begin{aligned}
^{237}_{93}\text{Np} \rightarrow\ &^{233}_{91}\text{Pa} + \alpha\\
&\hookrightarrow\ ^{233}_{92}\text{U} + e^- + \bar{\nu}_e\\
&\quad\ \hookrightarrow\ ^{229}_{90}\text{Th} + \alpha\\
&\qquad\ \hookrightarrow\ ^{225}_{88}\text{Ra} + \alpha\\
&\qquad\quad\ \hookrightarrow\ ^{225}_{89}\text{Ac} + e^- + \bar{\nu}_e\\
&\qquad\qquad\ \hookrightarrow\ ^{221}_{87}\text{Fr} + \alpha\\
&\qquad\qquad\quad\ \hookrightarrow\ ^{217}_{85}\text{At} + \alpha\\
&\qquad\qquad\qquad\ \hookrightarrow\ ^{213}_{83}\text{Bi} + \alpha\\
&\qquad\qquad\qquad\quad\ \hookrightarrow\ ^{213}_{84}\text{Po} + e^- + \bar{\nu}_e\\
&\qquad\qquad\qquad\qquad\ \hookrightarrow\ ^{209}_{82}\text{Pb} + \alpha\\
&\qquad\qquad\qquad\qquad\quad\ \hookrightarrow\ ^{209}_{83}\text{Bi} + e^- + \bar{\nu}_e
\end{aligned}
$$

where $^{209}_{83}\text{Bi}$ is stable.

30.4.4 Energy Released in Radioactive Decay

The spontaneous processes of alpha, beta, and gamma decay will not occur unless the sum of the masses of the product particles is less than the mass of the original atom. The difference in the mass times c^2 is the energy released.

Example

An example of an alpha emission is that of Radon-222, which decays to Polonium-218, with a half-life of approximately 3.8 days.

$$^{222}_{86}\text{Rn} \rightarrow {}^{218}_{84}\text{Po} + \alpha$$

Calculate the energy released in that decay.

Solution

Alpha decay results in the ejection of a helium nucleus from the nucleus of the radon. The atomic masses of radon-222, polonium-218, and helium-4 are 222.017 570 u, 218.008 973 u, and 4.002 603 u, respectively.

$$\Delta m = 222.017570 \text{ u} - (218.008973 \text{ u} + 4.002603 \text{ u}) = 0.005994 \text{ u}$$
$$E = \Delta mc^2 = [(0.005\ 994\ u)\ (1.66054 \times 10^{-27}\ \text{kg/u})]\ (2.9979 \times 108\ \text{m/s})^2$$
$$E = 8.9454 \times 10^{-13}\ \text{J} = 8.9454 \times 10^{-13}\ \text{J}\ \{1\text{eV}/[(1.6022 \times 10^{-19}\text{C})\ (\text{J/C})]\}$$
$$= 5583198\ \text{eV}$$
$$E = 5.5832\ \text{MeV}$$

Because the mass of the nucleus is so much greater than the mass of the alpha particle, almost all of the energy resulting from the decay is in the form of the kinetic energy of the alpha particle, but the nucleus does recoil; hence, a small fraction of the energy is in the form of the kinetic energy of the polonium ion.

There is an interesting exchange of electrons in the data that you should be aware of even though you don't have to treat them explicitly in the calculations. The atomic masses found in tables and provided in this example are for atoms with their full complement of electrons. It should be made clear that radon-222 has 86 electrons, so when it ejects an alpha particle to become polonium-218, it has two extra electrons, which are not included in the mass listed for polonium-218. This would be a problem except that the mass of the helium used for the alpha particle has two electrons not possessed by the alpha particle. This balances the masses. This also works for a beta decay in which an ejected electron compensates for the missing electron in the daughter atom, so the energy released is simply the energy associated with the mass difference of the isotopes involved.

30.5 ANSWER TO CHAPTER QUESTION

The chapter question is "Does exposing food to radiation make the food radioactive?" The answer to the question is no, radiation and radioactivity are different. Radiation describes the e&m waves that make up all parts of the electromagnetic spectrum, from radio waves to gamma rays. So, when you put your food in the microwave oven, it is exposed to radiation, and that will not make the food radioactive. Radioactivity, on the other hand, is a natural process by which an unstable nucleus in an atom ejects a particle and/or energy. As mentioned in the chapter, the gamma ray is a form of radiation that occurs in some radioactive process, but radiation does not make food radioactive.

30.6 QUESTIONS AND PROBLEMS

30.6.1 MULTIPLE CHOICE QUESTIONS

1. The naturally occurring isotopes of carbon are carbon-12, carbon-13, and carbon-14. On the periodic table, carbon has an atomic mass of 12.011 u, which tells us that carbon-12:
   A. is actually a little more massive than originally thought.
   B. has fewer protons than carbon-13 and carbon-14.
   C. is more abundant on earth than carbon-13 and carbon-14.
   D. has more neutrons than carbon-13 and carbon-14.

2. As time progresses, the decay rate of a radioactive sample decreases, because:
   A. the half-life is proportional to the mass of the sample of the sample.
   B. the remaining nuclei have lower probabilities of decaying.
   C. there is less of the radioactive nuclei remaining in the sample.
   D. all of the above are true.

3. In a single decay event, carbon-14 $\left({}^{14}_{6}C\right)$ decays to nitrogen-14 $\left({}^{14}_{7}N\right)$. What type of radioactive particle does carbon-14 produce?
   A. α (alpha)
   B. β (beta)
   C. γ (gamma)

4. Carbon-14 is a radioactive isotope of carbon that forms in the atmosphere when cosmic rays strike air molecules. It is taken in by living organisms as they respire, but when they die, the intake of carbon stops, so the ratio of carbon-14 to carbon-12 decreases, since C-12 is not radioactive. In a living organism, like a plant, the radioactive count rate of the carbon-14 is approximately 15 counts per gram per minute. The decrease in the ratio of C-14 to C-12, which is directly proportional to a decrease in the radioactive count rate, is used in carbon dating, since it is known that carbon-14 $\left({}_{6}C^{14}\right)$ decays into nitrogen-14 $\left({}_{7}N^{14}\right)$ with a half-life of approximately 5730 years. A sample of cloth (made from natural fibers) is taken from an archeological dig, which has a radioactive count rate of 3.72 counts per minute. How long ago was the plant material harvested and made into cloth?
   A. 5,730 years
   B. 11,460 years
   C. 17,190 years
   D. 22,920 years
   E. 28,650 years
   F. 34,380 years

5. A sample of Iodine-131 $\left({}^{131}I\right)$ initially has a count rate of 80 counts per minute at $t = 0$ days. If after 24 days the count rate is 10 counts per minute, what is the half-life of Iodine-131?
   A. 2 days
   B. 3 days
   C. 4 days
   D. 6 days
   E. 8 days
   F. 12 days

6. Uranium-238 $\left({}^{238}U\right)$ is radioactive, and it decays into Thorium-234 $\left({}^{234}Th\right)$. What type of radioactive particle does Uranium-238 produce?
   A. α (alpha)
   B. β (beta)
   C. γ (gamma)

7. Two pure samples of Cobalt-60 are produced on the same day, one with a count rate of 200 cps and another with a count rate of 400 cps. Is the half-life of the one with 400 cps greater than, less than, or equal to the half-life of the original 200 cps sample?
   A. greater than
   B. less than
   C. equal to

8. Bismuth-212* $\left(^{212}_{83}\text{Bi}^*\right)$ decays into bismuth-212 $\left(^{212}_{83}\text{Bi}\right)$ through what kind of radioactivity?
   A. α (alpha)
   B. β (beta)
   C. γ (gamma)

9. Bismuth-212 $\left(^{212}_{83}\text{Bi}\right)$ decays into polonium-212 $\left(^{212}_{84}\text{Po}\right)$ through what kind of radioactivity?
   A. α (alpha)
   B. β (beta)
   C. γ (gamma)

10. Bismuth-212 $\left(^{212}_{83}\text{Bi}\right)$ decays into thallium-208 $\left(^{208}_{81}\text{Tl}\right)$ through what kind of radioactivity?
   A. α (alpha)
   B. β (beta)
   C. γ (gamma)

30.6.2 PROBLEMS

Helpful Information: the atomic mass unit is $u = 1.6605 \times 10^{-27}$ kg

1. 100 mCi of ^{198}Au (which has a half-life of 2.7 days) is given to a patient for cancer therapy. If none is eliminated biologically, how much is left in 2 weeks?

2. ^{3}H (tritium) is one radioisotope used for whole body scans. Tritium has a half-life of 12.32 years. When the place of one of the hydrogen atoms in a water molecule is taken by tritium, the water is referred to as heavy water. Suppose that a sealed bottle of water is placed on a shelf at a time when it has .00300 moles of tritium in it. Assuming no leakage, 20 years later, how many moles of tritium are left in the bottle? *Hint: Use moles as the unit for N and N_o.*

3. ^{3}H (tritium) is one radioisotope used for whole body scans. Its physical half-life is 12.3 years, and its biological half-life is 19 days. What is its effective half-life? *Hint: Find the decay constant of each half-life and add them to find the total decay constant. Then, use that to answer the problem. This is how biological processes like respiration, perspiration, and other "-ations" are added into the calculation of how long a patient has from the time of ingestion to the time in which the body scan must be conducted.*

4. In β⁻ decay, the energy given off is in the form of kinetic energy and it is, in general, shared by the electron and the electron antineutrino. Assuming the mass of an antineutrino is negligible compared to that of an electron antineutrino so its kinetic energy is negligible, find the maximum possible kinetic energy of the electron emitted in the reaction in which lead-212 transmutes to bismuth-212 by beta decay:

$$^{212}_{82}\text{Pb} \rightarrow {}^{212}_{83}\text{Bi}^{1+} + e^- + \bar{\nu}_e$$

given that the mass of a neutral atom of lead-212 is 211.9918975 u, the mass of a neutral atom of bismuth-212 is 211.9912857 u, and the rest mass of an electron: $m_e = 9.1095 \times 10^{-31}$ kg $= 5.486 \times 10^{-4}$ u.

5. Cobalt-60 ($_{27}Co^{60}$) is a radioactive isotope of Cobalt that decays, with a half-life of 5.2174 years, into the stable isotope of Nickel-60 ($_{28}Ni^{60}$). A 1 g sample of Cobalt-60 is prepared for use in radiation therapy, and the count rate of the sample was measured with a Geiger counter at 96 counts per second (csp). After 20 years, what will the count rate of the Cobalt-60 sample be?

6. Calculate the energy released in a beta decay of Cobalt-60 to Nickel-60, $[_{27}Co^{60} \rightarrow \beta + _{28}Ni^{60}]$. The rest mass of Cobalt-60 is $m_{Co60} = 59.935348$ u, the rest mass of Nickel 60: $m_{Ni60} = 59.9307864$ u, and the rest mass of an electron: $m_e = 5.486 \times 10^{-4}$ u $= 9.1095 \times 10^{-31}$ kg.

7. Iodine-131 ($_{53}I^{131}$) is a radioactive isotope, which is used to destroy thyroid tissue in the treatment of an overactive thyroid. This isotope of Iodine ($_{53}I^{131}$) is a Beta (β) emitter with a half-life of 8.07 days. If a hospital receives a shipment of Iodine-131 with a count rate of 200 cps, what will be the count rate of the Iodine-131 in 30 days (approximately a month)?

8. Yttrium-90 ($_{39}Y^{90}$) is a radioactive isotope that is used in the treatment of some cancers and arthritis. This isotope of Yttrium ($_{39}Y^{90}$) decays to into Zirconium-90 (Zr) with a half-life of 2.67 days. If a hospital receives a shipment of Yttrium-90 with a count rate of 400 cps, what will be the count rate of the Yttrium-90 in 10 days?

9. Clinical trials are underway to treat leukemia and melanoma with targeted alpha therapy (TAT) employing a radioactive isotope of Bismuth ($_{83}Bi^{213}$), which is an α-emitter with a half-life of 61 minutes. If the initial count rate of a sample of pure Bismuth-213 is 2000 cps, what will be the count rate (cps) of the same sample in 7.5 hrs.?

10. Carbon-14 is a radioactive isotope of carbon that forms in the atmosphere when cosmic rays strike air molecules. It is taken in by living organisms as they respire, but when they die the intake of carbon stops. Since Carbon-12 is not radioactive, the ratio of Carbon-14 to Carbon-12 decreases after death. In a living organism, like a plant, the radioactive count rate of the Carbon-14 is 80 counts per minute per gram. The decrease in the ratio of C-14 to C-12, which is proportional to a decrease in the radioactive count rate, is used in carbon dating, since it is known that Carbon-14 ($_6C^{14}$) decays into Nitrogen-14 ($_7N^{14}$) with a half-life of approximately 6,000 years. A 1 g sample of cloth (made from natural fibers of a plant) is taken from an archeological dig. The radioactive count rate of the cloth is 3 counts per minute. How long ago was the plant material harvested and made into cloth?

Index